RADICAL
ABUNDANCE

RADICAL ABUNDANCE

HOW A REVOLUTION
IN NANOTECHNOLOGY WILL
CHANGE CIVILIZATION

K. ERIC DREXLER

PublicAffairs

New York

Published in the United States by PublicAffairs™,
a Member of the Perseus Books Group

PublicAffairs books are available at special discounts for bulk purchases in the U.S. by
corporations, institutions, and other organizations. For more information, please
contact the Special Markets Department at the Perseus Books Group, 2300 Chestnut
Street, Suite 200, Philadelphia, PA 19103, call (800) 810-4145, ext. 5000, or e-mail
special.markets@perseusbooks.com.

Book Design by Pauline Brown
Typeset in 11.5 point Minion Pro by the Perseus Books Group

Library of Congress Cataloging-in-Publication Data
Drexler, K. Eric.
Radical abundance : how a revolution in nanotechnology will change civilization /
K. Eric Drexler. — First
edition.
    pages cm
Includes bibliographical references and index.
ISBN 978-1-61039-113-9 (hardcover : alkaline paper) — ISBN 978-1-61039-114-6
(e-book) 1. Nanotechnology. 2. Nanotechnology—Social aspects. 3. Technology
and civilization. 4. Technological forecasting. 5. Social prediction. I. Title.
T174.7.D745 2013
303.48'3—dc23
                                                                            2012049031

First Edition

10 9 8 7 6 5 4 3 2

For my friend and adviser
Arthur Kantrowitz,
who this year would be 100

CONTENTS

A Necessary Prelude   *ix*

PART 1   AN UNEXPECTED WORLD

CHAPTER 1   Atoms, Bits, and Radical Abundance   3

CHAPTER 2   An Early Journey of Ideas   9

CHAPTER 3   From Molecules to Nanosystems   22

PART 2   THE REVOLUTION IN CONTEXT

CHAPTER 4   Three Revolutions, and a Fourth   39

CHAPTER 5   The Look and Feel of the Nanoscale World   55

CHAPTER 6   The Ways We Make Things   72

PART 3   EXPLORING DEEP TECHNOLOGY

CHAPTER 7   Science and the Timeless Landscape
of Technology   89

CHAPTER 8   The Clashing Concerns of Engineering
                 and Science   107

CHAPTER 9   Exploring the Potential of Technology   132

PART 4   THE TECHNOLOGY OF RADICAL ABUNDANCE

CHAPTER 10   The Machinery of Radical Abundance   147

CHAPTER 11   The Products of Radical Abundance   159

PART 5   THE TRAJECTORY OF TECHNOLOGY

CHAPTER 12   Today's Technologies of Atomic Precision   177

CHAPTER 13   A Funny Thing Happened on the
                 Way to the Future . . .   194

CHAPTER 14   How to Accelerate Progress   213

PART 6   BENDING THE ARC OF THE FUTURE

CHAPTER 15   Transforming the Material Basis
                 of Civilization   223

CHAPTER 16   Managing a Catastrophic Success   240

CHAPTER 17   Security for an Unconventional Future   259

CHAPTER 18   Changing Our Conversation About   273
                 the Future

                 *Appendix I: The Molecular-Level Physical Principles
                 of Atomically Precise Manufacturing*   289

                 Appendix II: Incremental Paths to APM   294

                 Acknowledgments   313

                 Notes   315

                 Index   329

A NECESSARY PRELUDE

IMAGINE WHAT THE WORLD might be like if we were *really* good at making things—better things—cleanly, inexpensively, and on a global scale. What if ultra-efficient solar arrays cost no more to make than cardboard and aluminum foil and laptop supercomputers cost about the same? Now add ultra-efficient vehicles, lighting, and the entire behind-the-scenes infrastructure of an industrial civilization, all made at low cost and delivered and operated with a zero carbon footprint.

If we were *that* good at making things, the global prospect would be, not scarcity, but unprecedented abundance—radical, transformative, and sustainable abundance. We would be able to produce radically more of what people want and at a radically lower cost—in every sense of the word, both economic and environmental.

This isn't the future most people expect. Over recent decades the world has been sliding toward a seemingly inevitable collision between economic development and global limits. As nations expand industrial capacity, carbon emissions rise. Expectations of resource scarcity drive wars and preparations for war as tensions grow over water from rivers, metals from Africa, oil from the Middle East, and fresh oil fields beneath the South China Sea. Everywhere progress and growth are beginning to resemble zero-sum games. The familiar, expected future of scarcity and conflict looks bleak.

These familiar expectations assume that the technology we use to produce things will remain little changed. But what if industrial production as we know it can be changed beyond recognition or replaced outright? The consequences would change almost everything else, and this new industrial revolution is visible on the horizon.

Imagine a world where the gadgets and goods that run our society are produced not in a far-flung supply chain of industrial facilities, but in compact, even desktop-scale, machines. Imagine replacing an enormous automobile factory and all of its multi-million dollar equipment with a garage-sized facility that can assemble cars from inexpensive, microscopic parts, with production times measured in minutes. Then imagine that the technologies that can make these visions real are emerging—under many names, behind the scenes, with a long road still ahead, yet moving surprisingly fast.

———

IN 1986 I INTRODUCED the world to the now well-known concept of nanotechnology, a prospective technology with two key features: *manufacturing using machinery based on nanoscale devices,* and *products built with atomic precision.* These features are closely linked, because atomically precise manufacturing relies on nanoscale devices and will also provide a way to build them.*

Nanoscale parts and atomic precision together enable atomically precise manufacturing (APM), and through this technology will open the door to extraordinary improvements in the cost, range, and performance of products. The range extends beyond the whole of modern physical technology, spanning ultra-light structures for aircraft, billion-core laptop

---

* If we were to stretch nanometers to centimeters (magnifying by a factor of ten million), atoms would look like small beads, nanoscale gears would look like gears with a beaded texture, and an electric motor could be held in the palm of your hand. As we'll see in Chapter 5, this magnified view—with time scaled in equal proportion—offers a surprising degree of accurate and yet intuitive physical insight.

computers, and microscopic devices for medical use, including devices able to recognize and destroy cancer cells.

Nanotechnology meant a profound revolution in production and products, and soon after 1986 the concept of nanotechnology took on a life of its own. In equal measure it sparked excitement and controversy, suggesting new research paths to the scientific community and exciting (if sometimes fantastic) futuristic visions to our popular culture. The idea of building things on the molecular level soon spurred the growth of fields of research; a decade later, these fields had grown into billion-dollar programs around the world, all devoted to studies of nanotechnology.

During the 1990s, however, public and scientific visions drifted and followed divergent paths. The futuristic popular visions floated free from reality, into realms unconnected to science, while the scientists themselves turned toward work that would bring in research funds, with a focus on short-term results. As popular expectations skewed one way and research in another, what was called "nanotechnology" began to seem like a hyped disappointment—a broken promise, not an emerging revolution that would reshape our world.

In recent years technology has advanced surprisingly far toward a critical threshold, a turning point on the road to APM-level technologies. While progress in atomically precise fabrication has accelerated, understanding of its implications has lagged, not only in the public at large, but also within the key research communities. Much of the most important research is seldom called "nanotechnology," and this simple problem of labeling has obscured how far we have come.

Understanding matters and ignorance can be dangerous. The advent of a revolution in nanotechnology will bring capabilities that transform our world, and not in a small way. The ramifications encompass concerns on the scale of climate change, global economic development, and the gathering crises of the twenty-first century.

The revolutionary concept is simple in essence, as such things often are.

The key is to apply atomically precise nanotechnologies to build the machines we use to make things. Large scale, high-throughput atomically precise manufacturing is the heart of advanced nanotechnology, and in the coming years it has the potential to transform our world.

APM is a kind of manufacturing, but it isn't *industrial* manufacturing. The differences run from bottom to top and involve replacing enormous, polluting factories with clean, compact machines that can make better products with more frugal use of energy and material resources.

The Industrial and Information Revolutions can serve as models (and yardsticks) because atomically precise manufacturing will combine and amplify the features of both. What computer systems have done for processing information, APM systems will do for processing matter, providing programmable machines that are fast, inexpensive, and enormously flexible—like computers in many ways, but rather than electronic signals, producing physical products.

Rough as it may be, the comparison to computing is useful because APM has much in common with digital electronics. The parallels range from their shared basis in fast, discrete operations to their emergent similarities in scale, speed, cost, and scope of application. Where digital electronics deals with patterns of bits, APM deals with patterns of atoms. Where digital electronics relies on nanoscale circuits, APM relies on nanoscale machinery. Where the digital revolution opened the door to a radical abundance of information products, the APM revolution will open the door to a radical abundance of physical products, and with this, a cascade of transformative consequences that history suggests will amount to a Version 2.0 of world civilization, a change as profound as the Industrial Revolution, but unfolding at Internet speeds.

As progress accelerates toward the APM revolution, we as a society would be well advised to devote urgent and sober attention to the changes that lie ahead, taking account of what can be known and the limits of knowledge as well. At the moment, however, even the basic facts about this kind of technology have been obscured by confusion and science-fiction fantasies.

Imagine standing back in the late 1960s and looking forward to prospects for microcomputers based on progress in microelectronics. Now imagine that the public had somehow confused micro*electronics* with micro*biology*, and expected microbes to compute, or chips to produce insulin. Now stir in popular fantasies about genetic engineering, bizarre mutants, and armies of clones. . . .

Micro-this, micro-that—how much difference can there be between one kind of tiny thing and another? The answer, of course, is "almost everything." Rocks, dogs, lawnmowers, and computers have little in common beyond meter-scale size, and things measured in microns or nanometers are just as diverse.

This imaginary history of confusion about microelectronics has all too much in common with the actual confusion that enveloped nano-technology, a confusion that emerged in the late 1980s.

The public isn't to blame for this confusion. In the past decades the concept of nanotechnology itself has been stretched almost beyond recognition to embrace a wide range of nanoscale technologies. In Washington the promoters of a federal nanotechnology program sold a broad initiative to Congress in 2000 and then promptly redefined its mission to exclude the molecular sciences, the fields that comprise the very core of progress in atomic precision. Thus, the word "nanotechnology" had been redefined to omit (and in practice, exclude) what matters most to achieving the vision that launched the field.

Now imagine the press trying to untangle this story and convey it to an already bewildered public. It just didn't happen. The resulting muddle has obscured both the nature of the critical technologies and the pace of progress along paths that lead to APM. Many readers will be surprised to learn how far we have come and how close we really are.

It's time to put years of nonsense behind us and start afresh.

Through this book I invite you on a journey of ideas that begins with common knowledge, yet leads in uncommon directions. This journey traverses a landscape of concepts with APM in the center and offers views of that center from perspectives that range from scientific and technological to cultural, historical, cognitive, and organizational. Along the way we will climb toward a vantage point that offers a glimpse of a better future and what must be done to get us there.

In the end my aim is not to convince, but to raise urgent questions; not to persuade readers to upend their views of the world, but to show how the future may diverge far from the usual expectations—to open a staggering range of questions, to offer at least a few clear answers, and to help launch a long-delayed conversation about the shape of our future.

NOTE, OCTOBER 2012

Just over thirty years ago, I worked with a typewriter keyboard to outline a path toward a general-purpose, atomically precise fabrication technology; in 1981, the resulting paper saw print in the *Proceedings of the National Academy of Sciences* and launched fruitful research efforts along the lines I'd proposed.

The following year I worked with a computer keyboard to describe prospects for that atomically precise fabrication technology, a concept I called "nanotechnology"; in 1986, after many revisions, the resulting book reached the public and events snowballed from there.

Today, I work with a different computer, a machine with ten thousand times more processor power, one hundred thousand times more memory, and one million times more disk capacity—a set of advances enabled by devices built at a nearly atomic scale.

Within this same span of time (yet beyond eyesight or touch), the scale of true atomically precise fabrication has grown from building structures with hundreds of atoms to building with millions. The pathway technologies that I outlined in 1981 are now approaching a threshold of maturity, a point of departure for yet faster progress.

We've come a long way along a path that leads toward a highway and it's time to count up the milestones, read the signposts, and look forward.

PART 1

AN UNEXPECTED WORLD

CHAPTER 1

Atoms, Bits, and Radical Abundance

New ways to put parts together can transform broad realms of human life. We've seen it happen before and it will happen again, sooner than most people expect.

═══

NOT SO LONG AGO, if you wanted to bring the sound of a violin into your home, you would have needed a violin and a violinist to play the instrument. For the sound of a cello to accompany the violin, you would have needed a cello and a cellist, and to add the sound of a flute, you would have needed a flute and a flautist. And if you wanted to bring the sound of a symphony orchestra into your home, you would have needed a palace and the wealth of a king.

Today, a small box can fill a room in your home with the sound of a violin or of a symphony orchestra—drawing on a library of sound to provide symphony and song in radical abundance, an abundance of music delivered by a very different kind of instrument.

Looking back, we can see a radical break that divides the past from the present. Behind each violin stood a craftsman, a link in a chain spanning generations, each refining the previous generation's instruments of hand-crafted sound. Behind each of our modern machines, in contrast, stands a new global industry that creates music machines without any link to the traditions of resonant wood, string, rosin, and bow. Each of today's machines instead contains silicon chips, each bearing a host of nanoscale digital devices spread across its surface—millions, even billions of transistors linked by strips of metal to form intricate electronic circuits.

═══

NOT SO LONG AGO, in order to print words on a sheet of paper you had to arrange pieces of metal, each in the shape of one of the letters and found in a tray full of type. If you fancied changing the font or typeface of the letters, you would have had to take different pieces of metals from a different tray. To print pictures, you would use engraved metal plates, and to print a page using these pieces of metal you would need ink and a machine to press the inked metal against paper. A single print job could require hours and days of tedious work. Printing would have been beyond practical reach without a print shop, customers, and income to pay a team of assistants to keep the press running.

Today, affordable desktop machines can print any pattern of letters and images without the need for a print shop, customers, or skilled labor, producing a radical increase in access through a radically different kind of machine.

Just as with music and violin-making, printing was a craft transmitted through a chain of apprentices. And once again, in the world today there is a new industry based on machines that embody a radical break from previous crafts, and at the heart of each modern printing machine, a host of nanoscale digital devices on silicon chips.

═══

NOT SO LONG AGO, when I was in school, research required a trip to a library stocked with bundles of printed paper—an inconvenient undertaking when the nearest good library was miles away. Today, affordable machines can deliver the content of a library's journals to your lap in an instant—and behind this modern wonder we again find silicon chips with digital devices.

Mail that arrives in an instant, not carried by trucks or delivered by hand? Movies at home that arrive in an instant, without film or a theater? Conversations with friends thousands of miles away, without wizards or magic? Once again, at the heart of it all, we find silicon chips bearing nanoscale digital devices, the electronic machinery that transmits text, paints movies on screens, and delivers voices to telephones.

Each of these developments carries a double surprise. First, from the perspective of pre-industrial times, is the surprise of their very existence. A second, more profound surprise, however, is how they work, in the most basic sense, their unified technological basis and its radical scope.

Imagine yourself in pre-industrial times and consider how implausible each of these recent advances would have seemed. To an artisan skilled in the crafting of violins, an iPod would seem frankly preposterous. To a worker in a print shop in the seventeenth century, the power and outward simplicity of a desktop printer would be beyond imagining.

Now place yourself in the mid-twentieth century, just before the digital revolution took hold. By that time, each of these capabilities would have seemed possible—indeed, most already existed, though enabled by different technologies:

Music-makers without musical instruments—Phonographs.
Printers without pieces of metal type—Offset lithography.
Instant mail across miles—Telegraphs and teletype machines.
Transoceanic conversations—Cables and telephones.
Movies at home—Movie projectors.

And a library's journals, available on demand? In the closing months of World War II, Vannevar Bush proposed a desk-scale machine to retrieve images of pages stored on microfilm. If such a machine had been built to hold data on a library scale, however, its cost would have been enormous.

For each of these capabilities, then, the conceptual sticking point wasn't the ends, but the means; not the idea of broad progress, but the form this progress would take and how far-reaching it would be. Surely, in light of the whole history of engineering, an advanced music player would be simply a sound-making machine, not also a teletype, a library, and a movie projector—and surely not also a typewriter, drafting table, calculator, filing cabinet, and photo album, too, and a camera, a case-load of film, and a fully-equipped darkroom—and certainly not all of these devices somehow jammed together into a single box.

Yet with just one substitution (in place of a printer, a screen) the machine under my fingers can perform every one of these functions. This is what would have astounded an engineer of the mid-twentieth century: The extreme generality of the underlying, digital mechanisms and of the machines that can be built using this kind of technology.

Progress proceeded along more traditional lines until the digital revolution took hold. Explosive advances in digital information systems, combined with what became known as peripheral devices, changed the course of our technology, economy, and culture.

Every single part of these systems works on the same basic principle, creating complex patterns from small, simple parts—slicing sound into samples, images into pixels—and representing each part by means of patterns of bits that are then processed by arrays of small, simple nanoscale devices—transistors that implement the bit-by-bit information processing that defines digital electronic systems.

Building devices with components of nanoscale size it became possible to fit billions of transistors on a single chip and to work at gigahertz frequencies. The chips are products of a particular kind of nanotechnology, delivered by a specialized physical technology that produces general-purpose information machines.

In this limited sense, a nanoscale technology revolution has already arrived, bringing with it the radical abundance we call the Information Revolution. We haven't seen the end of this kind of revolution, however. The same profound digital principles will enable a parallel revolution that will enable radical abundance, not just in the world of information, but in the world of tangible, physical products as well.

FROM THE INFORMATION REVOLUTION TO APM

What digital technologies did for information, sound, and images, atomically precise manufacturing (APM) can do for physical products. This assertion raises a host of questions, but first, the parallels.

Consider this description of digital technologies:

Digital *information* processing technologies employ nanoscale *electronic* devices that operate at high frequencies and produce patterns of *bits*.

With a change of tense and a few words replaced, the same description applies to APM-based technologies:

APM-based *materials* processing technologies will employ nanoscale *mechanical* devices that operate at high frequencies and produce patterns of *atoms*.

As a first approximation, think of the process of forming a molecular bond as a discrete operation, i.e., all or nothing, like setting a bit in a byte to a 1 or 0, and think of an APM system as a kind of a printer that builds objects out of patterns of atoms just as a printer builds images out of patterns of ink, constrained by a limited gamut, not of colors, but of output materials. Although the products are made with atomic precision (every atom in its proper place), this does not entail moving individual atoms. (From the standpoint of chemistry, recall that, by definition, regio- and stereo-specific reactions of molecules yield specific patterns of atoms, and do this without juggling atoms one at a time.)

The parallel between APM and digital information processing extends to the underlying physics as well, because they both rely on noise margins to achieve precise, reliable results. Noise margins in engineering allow for small distortions in inputs much as a funnel can guide a slightly misplaced ball through a selected hole in the top of a box. In mechanically guided chemical processes, elastic restraints on motion paths in effect guide bound molecules toward their intended targets. Thus, elastic restraints function as barriers, and for well-chosen reactions higher barriers can suppress thermally induced misplacement errors by a large, exponential factor. What this means is that in both nanoelectronic and nanomechanical systems, noise margins can be engineered to exceed the magnitude of disturbances and can suppress errors down to far less than one in a trillion.

As with today's digital systems, the potential power of APM results from an ability to produce complex patterns from their simplest parts. In much the same way that a music machine produces (within broad limits) any pattern of sound and a printing machine produces (within broad limits) any pattern of ink, APM-based production systems will be able to produce (within broad limits) any pattern of materials, and hence an extraordinary range of physical artifacts.

There's a crucial difference, however.

Audio systems produce complex patterns of sound, but our world isn't made of sound.

Printing systems produce complex patterns of ink, but our world isn't made of ink.

APM-based production systems, by contrast, will be able to produce patterns of matter, the stuff of audio systems, printers, production systems, and everything else that we manufacture, and more.

Perhaps even a violin.

———

AT THIS POINT, I imagine readers asking a natural question: If APM is a realistic prospect, why isn't it already familiar? This question is best understood through a history of the relevant ideas, science, and politics, interwoven with an exploration of the technology itself. Understanding the past can help us judge the state of the world today, and then survey the prospects for an unexpected future.

CHAPTER 2

An Early Journey of Ideas

THE STORY OF NANOTECHNOLOGY stretches back more than twenty-five years and is a fabric composed of many threads, woven of science, technology, myths, achievement, delay, money, and politics. It includes the rise of ideas and their collision with popular culture, the rise of lines of research and their collision with fashions in science, together with promises made, promises broken, and the emergence of a renewed sense of direction. I've been in the midst of this story from the very beginning.

Nanotechnology's promise, both real and imagined, has been shaped by its past, so to understand today's choices and challenges we must begin with a look back. The story begins with the discovery of what known physical law implies for the potential of future technologies.

In outline, the story had a simple beginning. The concepts that launched the field of nanotechnology first appeared in recognizable form in a scientific paper I published in 1981.* In that paper I described accessible paths in the field of atomically precise fabrication, paths that began with biomolecular engineering, and then went on to discuss the

---

* "Molecular Engineering: An Approach to the Development of General Capabilities for Molecular Manipulation," *Proceedings of the National Academy of Sciences USA,* 78, no. 9 (1981): 5275–5278.

fundamental principle of atomically precise manufacturing (APM): the use of nanoscale machines to guide the motion of reactive molecules in order to assemble large complex structures, including machines, with atomic precision. This concept, with its many applications, led directly to more.

In 1986, *Engines of Creation* brought a range of these concepts to the attention of the general public, describing and naming a vision I called "nanotechnology." Six years later, I updated and grounded this vision in a technical, book-length analysis based on my MIT dissertation, yet it was *Engines of Creation* that served as the flashpoint for all that followed.* The ideas I expressed drew worldwide attention and stirred a wave of excitement that launched (and then helped to fund) a field of research called "nanotechnology" in the years that followed. As the story unfolds, we will see how the initial vision and the emerging field intersected.

ON A MISSION THAT LED TO LIBRARIES

The path that led me to the concept of APM was a journey of ideas, driven by curiosity and guided by a sense of mission shaped by concerns at a world-wide scale that could be measured in terms of generations. That mission, as I first understood it, was to do my part to help save the world from a distant catastrophe, a collision of industrial civilization with the limits of the Earth itself. I saw my role as that of an explorer of potential technologies that could change the world situation, studying these technologies with the tools of engineering and science and then sharing what I had learned.

This sense of mission first gripped me in high school (a good time in life for grand dreams), and it coalesced in its first concrete form in 1970, the year of the first Earth Day.

I recall a bicycle ride, starting soon after dawn, on a forty-mile round-trip journey to an engineering library. The journey itself, often repeated

---

* *Engines of Creation: The Coming Era of Nanotechnology* (New York: Doubleday, 1986); and *Nanosystems: Molecular Machinery, Manufacturing, and Computation* (Hoboken, NJ: Wiley/Interscience, 1992).

that summer, reminded me of what was at stake. The path followed a road through the Oregon countryside, a road that climbed over hills flooded with sunlight. The trip through the summer heat brought a reminder of long forgotten forests. On the slope of a single hill stood a wooded patch, and down from its shadowed spaces poured cool, damp air that flowed across the road that I traveled.

Beyond the foot of the hill, farmland stretched across the Willamette Valley toward distant mountains. And beyond the horizon, yet visible in the mind's eye, the world was changing, year by year, as industrial growth drained resources, new arable lands became scarcer, and a growing population pressed against the elastic yet firm limits of Earth.

At the time, I thought I saw a potential way out. Keep in mind that these were times when men still walked on the Moon and dreams of settling distant planets were at their peak. It seemed to me, however, that the greatest potential for a future in space lay not on the surface of barren planets like Mars (places like Earth, but smaller and hostile to human habitation), but instead in the vast reaches of space itself, a sun-drenched realm of resources awaiting the touch of Earth's life, as the realm of Earth's continents had awaited the first touch of life from the sea.

This vision for the human future, which emerged from multiple sources, came to be known as "space development," and from the start prospects for space development raised questions that could be answered only by imagination shaped and disciplined by the study and analysis of quantifiable technological concepts.

In a world where computers rarely did more than compute, my search for knowledge and answers soon led to libraries, and truly useful libraries had to be large. A few miles from home, the Oregon College of Education's library stood open, yet it held few books on space science. The road across farmlands and hills, however, led to Oregon State University, where an open library embraced not only space science, but space systems engineering. At OSU, I found books that taught the principles of spacecraft engineering, a sample of the eternal physical principles that give all engineering its form.

For me, the concept of space development served not as a final destination, but as a kind of map. Space development would require new methods for manufacturing, while an understanding of what was possible there required studies of production methods suited to new environments. In an abstract sense, these studies of space development provided a roadmap for research when I turned from outer space to the nanoscale world.

Looking back across forty years of exploring ideas, I see a common direction. The same sense of mission guided my life's path throughout, turning first toward space, then toward advanced nanotechnologies, then, through a keyboard today, to share what I've learned, and how, and why.

Where had this sense of mission come from? In part, from broad social concerns about the future of industrial civilization and, in part, from a particular time in the history of science and technology. On the timeline of developments in molecular science and space technology, James Watson and Francis Crick in Great Britain had mapped the atoms of DNA just three years before I was born, while Sergei P. Korolev's engineering team in the Soviet Union had launched Sputnik 1 into space just two years after. My mother, Hazel, had clipped and saved newspaper reports of the first satellite launches because she thought I'd be interested, then fed me a diet of science fiction and science that helped that interest to grow.

This diet of books shaped my perspective, but it was the early environmental movement that infused me with a sense of mission. Along with books on space came a book on a sobering topic: the cumulative ecological consequences of spraying millions of acres of crops with tens of thousands of tons of organochlorine pesticides per year (which, the book noted, far exceeded the amounts needed for mosquito control), spreading poisons that persisted for years and accumulated in animal tissues, then passed from prey to predator, becoming more concentrated, more toxic at every step up the food chain. The book was Rachel Carson's 1962 bestseller, *Silent Spring,* widely credited with boosting the environmental movement past a tipping point. Late in the decade, Hazel read Carson's book and then passed it on to me. This kind of reading had its effect, and in April 1970 I joined others (in a minor, high-school way) in boosting the first Earth Day.

Two years later I encountered a book that changed my view of the world more profoundly: *Limits to Growth*.* The book undertook an audacious goal: to model the underlying dynamics of global growth as an interlinked process, assuming that technology, resources, and the environment's resilience would remain within plausible bounds. The models that were presented in *Limits to Growth* suggested that continued economic growth, at first following an exponential trend, would lead to disaster in the early to middle decades of the 21st century. Contrary to later critics' assertions, the authors claimed no predictive ability, but, more modestly, argued that such models provided "indications of the system's behavioral tendencies": growth, overshoot, and collapse. Changing the input parameters in different scenarios led to collisions with different limits or several together, but unconstrained growth always led to disaster.

In the years since, critics have attempted to dismiss *Limits*, often claiming that the book wrongly predicted collapse in the late twentieth century. It didn't—not even the worst-case scenarios gave that result. Instead, the book's baseline scenario for the early twenty-first century strongly resembles the world we see today.

At the time, the Malthusian message of *Limits to Growth* seemed more than plausible, and if taken seriously, seemed to nail a lid on the human future. To my eyes, however, every model in *Limits* shared a crucial defect: When the authors framed their models of world dynamics, they included only the Earth. That is to say, the authors had set aside as irrelevant almost the whole of the universe—and at a time when men still walked on the Moon and looked far beyond. At the time, NASA promised low-cost access to space. At the time, bold dreams flourished and the world beyond Earth seemed within practical reach.

The restricted vision embodied in *Limits to Growth* raised questions that led me to explore what might be found outside the world it had framed—to look outward, at first, toward deep space, but later inward, to explore the potential of technologies in the nanoscale world.

---

* Donella H. Meadows, *The Limits to Growth. A Report for the Club of Rome's Project on the Predicament of Mankind* (New York: Universe, 1972).

With the end of high school less than a year away, I applied to MIT. My grades weren't outstanding, but I tested well, and that proved to be enough.

At MIT I soon felt that I had come home; people understood what I said and filled gaps in my knowledge, and the libraries seemed endless.

At first, I found few who shared my view that free-space development had potential importance, while planetary surfaces were a distraction. The seeming lack of discussion of the subject gave me reason to doubt my previous confidence. Had I been mistaken about the promise of the space frontier? Or could it be that my better-informed elders had over-looked something important, that they had asked the wrong questions?

Indeed, I found that few had asked the right questions, and therefore few had considered and weighed the potential answers. Engineers and space planners, at NASA and elsewhere, had asked "How can we explore and survive on other planets—the places in space most like the Earth?" The question I asked was, "Where can we find an environment that can sustain a vibrant industrial civilization?" This different question had a different answer, and free-space development had no connection with distant planets.

In search of someone who shared this vision of the latent potential of the world beyond Earth, I asked my freshman adviser to direct me to someone who might know someone who knew something about this sort of idea. He did, and the professors he suggested both directed me to a revered MIT physicist, Philip Morrison. After the second recommendation, I gained the courage to knock on his door. He did indeed know something, and someone.

Professor Morrison directed me to a professor of physics at Princeton, Gerard K. O'Neill, who (as it happened) was planning a conference centered on his vision of space development. This vision had something in common with my own line of thought. It set planets aside in favor of space itself as a place for Earth's life, and it proposed ways to avoid a cataclysmic collision between human civilization and Earth's limits to growth. From there, however, his vision gave less weight to concerns with materials and manufacturing, and highlighted instead a concept that strongly engaged the public's imagination—a grand and very *visual* vision of new lands built in free space itself.* O'Neill had published calculations of geometry, light, atmospheric pressure, centripetal acceleration, and structural mass for vast cylindrical structures—kilometers in scale—all based on the known properties of sunlight, glass, mirrors, and suspension-bridge steel. These space habitats were to be open spaces large enough to hold cities and farms, places with sunlit lands, the feel of gravity underfoot, and, with the passage of years, forests. Perhaps most important of all, this concept inspired artists to portray visions of places in space that looked much like home, images that gripped the public imagination

As a freshman, I found myself playing a minor role in organizing the first Princeton conference on space colonization—a term NASA later amended to "space settlement" at the request of the State Department. As a result of this meeting, a community began to coalesce, an eclectic mix that ranged from undergraduates and scientists to aerospace systems engineers and environmental activists. Study groups and summer studies followed, together with reports, conference papers, press coverage, critics, and even a popular movement of sorts.

The vision of space settlement had deep resonance at a time when society had begun to question the material foundations of its own existence. A common concern about terrestrial limits to growth animated the space movement. Space is large, holding room and resources enough to open up realms for life on a scale of a thousand Earths. This physical potential suggested a path for civilization that could avert overshoot and catastrophe for centuries to come.** What's more, free-space development could lift the burden of industry from the biosphere and make room in the world for restoring the Earth.

---

* In fact, O'Neill imagined building with resources mined from a visible source that loomed large in the human imagination, the Moon, while I advocated using the more attractive resources offered by the less charismatic asteroids; the concept of asteroid mining at first gained little traction, yet missions to asteroids have become part of NASA's plans, now slated for 2025, before any return to the Moon.

** Not forever, of course, because in the end Malthus was right. Buying time for dozens of generations, however, seemed reason enough, and with this, perhaps time enough for humanity to gain wisdom enough to face future limits with a measure of grace. Stranger things have happened in the long arc of history.

RADICAL ABUNDANCE

The mid-1970s was the time of "the energy crisis," when OPEC-created oil shortages had highlighted the idea of terrestrial limits and thereby spurred a search for ways to transcend them. Engineers proposed that solar power beamed from space could compete with terrestrial sources of energy, and so NASA and the Department of Energy provided research funds to aerospace firms to support exploratory design and analysis of potential solar power satellite systems. The idea of building these massive satellites using resources already in space had appeal and soon became part of the space settlement concept.

This kind of large-scale construction would require space-based manufacturing, and a comprehensive infrastructure for space industrialization.

SCIENCE AND SPACE FOR MANUFACTURING

Manufacturing makes modern society possible. Food, clothing, shelter, travel, communications, and the conveniences of daily life—in the developed world all these rely on industrial products made by what are now increasingly automated processes. Societies in space would depend on industry to an even greater extent, in fact, for producing every bit of material, even soil and air. This is why the practical questions of space settlement quickly turned toward questions of mining, refining, and manufacturing.

At MIT I majored in a program called "Interdisciplinary Science," yet much of my study revolved around industry and agriculture—how they work on Earth and how their technologies could be recast to fit a radically different environment. To explore this area required understanding facts and principles from a wide range of fields. Some of these principles described macroscale phenomena, like the physics of heat and mass transfer; others led to the molecular world and the foundations of materials science. Yet other topics included meteoritics and planetary science, plant physiology and ecosystem engineering, photovoltaic cells and solar energy, vacuum metallurgy and the distillation of steel, the properties of materials that can withstand the incandescent temperatures of a solar furnace, and ways of stitching together terrestrial industrial processes to make glass and metals from lunar rock.

One line of study led me into the nanoscale world: designs of lightsails—rotating structures, kilometers wide, tiled with sheets of thin metal film, capable of harnessing the pressure of sunlight to drive vehicles through space, year after year, with a small but steady acceleration and no need for fuel.

Lightsails held my attention for several years (and a thesis) at MIT. Physical data showed that lightsails could catch and reflect sunlight using sheets of aluminum no more than 100 nanometers thick, or about 300 atomic diameters. Calculations and library research persuaded me that films of this thickness would serve their purpose if they could be made and installed in the vacuum of space, yet no calculation could persuade me that such delicate films could survive a manufacturing process. And so I learned to use vacuum equipment to deposit vaporized aluminum onto a surface, forming thin films, atom by atom. The films were indeed extremely delicate. In trying to free them, I tore apart one film after another. If freed and then touched, the thin metal film would mirror-coat a fingertip, wrapping each ridge in the skin and yet feeling like nothing at all. Set free in the air, a torn fragment of film would drift like a dust mote, yet reflect light like a scrap of aluminum foil. In the end, I learned to lift and mount these thin films in frames (and even took samples to Pasadena for a presentation at NASA's Jet Propulsion Laboratory), and through this hands-on experience I learned enough to conclude that automated machines in the space environment could indeed produce lightsail film in enormous quantities.

The method I learned came from the shelves of the MIT Science Library, while the science I learned showed me how things could be built from the bottom up, atom by atom.

INTERLUDE: ARTHUR KANTROWITZ

Early in those years the MIT Space Habitat Study Group grew out of a talk I gave on space settlement. Most members were students, but at a meeting one evening, a gray-haired man walked in and stayed in my life.

Arthur Kantrowitz was a physicist and engineer, an MIT Institute Professor (Visiting), the founder and head of the Avco Everett Research

Laboratory, and, I think, a wise man. Born in 1913, he was older then than I am today. He became my mentor and friend.

Over the years, Arthur shaped my view of the world, how it works, and what matters. He helped me understand the underlying nature of scientific knowledge and scientific norms, as well as the turbulent process that leads to new technologies. He shared his knowledge of the dirty side of that process, the secrecy and corruption that can flourish at the junction of policy, money, and technology. And beyond this, he shared his understanding of the underlying incentives and cultural problems, and his experience with attempts at institutional reforms.

As I look back, I can see how much of my sense of the world reflects his values.

Arthur had achieved bold and wide-ranging accomplishments in technology. In the 1950s, his inventions helped solve the problem of hypersonic atmospheric re-entry (the *New York Times* described him as "one of the first technological heroes of the space program"), yet in his youth practical aeronautics still centered on biplanes built of wood and cloth.

While leading research and development teams, Arthur pioneered a range of technologies that included high-power lasers, supersonic molecular beams, magnetohydrodynamic generators, and (with his brother, Adrian) the intra-aortic balloon pump, a heart-assist device now used in every major cardiac surgery center.

His bold visions started early. In 1939, with a colleague at what is now NASA's Langley Research Center, Arthur built the first laboratory machine to explore the potential of magnetically confined nuclear fusion power; in 1963, he reexamined the field and concluded that the entire approach faced a brick wall—nonlinear plasma instabilities—that to this day, a half century later, has not been surmounted. Arthur was bold and persistent, and he knew when to quit.

Arthur had experience with the space program from the inside and at high levels. At the inception of the Apollo program, for example, he served on a presidential commission that assessed the prospective costs, times, and development risks of competing approaches for reaching the Moon. In the 1970s, Arthur took a keen interest in the drive toward space development, giving talks, supporting organizations, leading research in high-capacity, small-payload space launch systems, and advising an MIT undergraduate who absorbed at least some of what he could teach.

It was Arthur who introduced me to the works of Karl Popper, the philosopher of science who established the principle that science can test ideas and sometimes approach the truth, yet can never prove a universally quantified theory. Popper called for an intellectual life of bold conjectures, tentatively held and subject to critical discussion and stringent testing. Grappling with Popper's view of epistemology (and with books by his critics) led me to a lifelong concern with the basis of knowledge in both science and engineering, and through this concern, to methodologies that have guided my life's work in exploring the potential of physical technologies.

Arthur was a man of both the future and the past. In a time of growing specialization, he was a generalist. In a time of growing timidity, he was bold. In a time of science increasingly driven by funding and politics, Arthur was a voice for the deeper values that make science work.

Because of Arthur, however, I misjudged the world. In a tacit, unconscious way, I assumed that science held many more people like him.

At the age of ninety-five, Arthur Kantrowitz died of a heart attack while visiting his family in New York. His last hours were good, I'm told, hours spent with his family while his life was sustained by a device he knew well, the intra-aortic balloon pump. I miss him more deeply than I would ever have guessed.

A CULTURE OF QUANTITATIVE DREAMS

My years of engagement with Arthur and others in the space systems community taught me a way of thinking that harnessed creative vision to physical, quantitative reasoning in order to explore what could be achieved in new domains of engineering.

The space systems engineering community has evolved together with the space systems themselves. Satellite launchers and moonships grew out of quantifiable engineering visions, system-level concepts that could be sketched, assessed, and discarded at a rapid pace, evolving through a kind of Darwinian competition. The best concepts would win the resources of time and attention needed to fill in more details, to optimize designs, to apply closer analysis, and after this refinement and testing, to compete again. The prize at the end would be a design refined into fully detailed specifications, then metal cut on a factory floor, then a pillar of fire rising into the sky bearing a vision made real.

For example, before President Kennedy committed the United States to the Apollo program, engineers had examined hundreds of ways of assembling rocket engines and fuel tanks to build systems that could reach the Moon. Much the same can be said of how ideas have evolved before every major space mission.

To play this game well requires creativity harnessed to skeptical evaluation, with attention to both knowledge and uncertainty. In a system-level engineering design—whether a sketch or a more detailed specification—every assumption, calculation, and uncertainty range can be critiqued. Uncertainties can be fatal or minor; some can be hedged, while others spur new research programs. In opening the space frontier, for example, one research program established how a spacecraft could survive a return to the Earth at hypersonic speeds through air heated to temperatures found on the Sun—the re-entry problem that Arthur addressed—thereby answering skeptics and squeezing engineering uncertainty into ranges narrow enough to enable more detailed and confident system design.

The space systems engineering culture shaped how I thought about problems at the junction of complexity, uncertainty, and exploratory design. It was from this milieu that I turned my attention to molecular technologies.

In those years, the space development community, supported by federal funding, explored decades-long plans for developing solar power satellites and space habitats, to be enabled by lower-cost successors to the yet-unbuilt Shuttle. Lightsails could play a role in that world, yet I found my attention drawn away in a different direction, toward the exploration of the potential of smaller and more complex things—not broad, nanometers-thick films of aluminum, but nanoscale devices and machines, the potential fruits of advances in molecular technologies.*

---

* The learning that prompted this line of thought came from reading journals like *Science* and *Nature*, and from dipping into more specialized journals, such as *Angewandte Chemie* (all of which I found, of course, on MIT's library shelves).

Once again, in the molecular sciences, it seemed to me that the experts were focused on different questions than those of the greatest long-term importance, that they were too close to their work to see where their fields could lead, if combined and applied to new ends. And as with exploring the potential of space, the questions and answers once again involved system-level engineering principles, and exploring how one might make things in an unfamiliar world. And once again, I found implications for the human future on a scale too large to ignore.

The same sense of mission that led me to explore the potential of space now pulled me toward the molecular world. The scientific knowledge already in place was enormous, and growing.

CHAPTER 3

From Molecules
to Nanosystems

THE IDEA OF BUILDING THINGS with atomic precision often strikes people as futuristic, yet atomically precise fabrication has a longer history than spaceflight, or even wooden biplanes. The story of atomically precise fabrication begins more than a century ago, at the start of an arc of accelerating progress.

By 1899, chemists had gained skill in building structures with atomic precision, and they knew what they were doing well enough to draw correct diagrams of molecules, atom by atom and bond by bond. Chemists knew, for example, that carbon atoms form four bonds, typically directed toward the corners of a tetrahedron, and that molecules therefore can have chirality, that is, they can have both left- and right-handed forms. They knew that carbon can form double bonds and that benzene consists of a ring of six equivalent carbon atoms, and they had also inferred how methyl groups could be attached to those rings in the patterns that define different molecular isomers. This was a remarkable degree of knowledge, considering that no one could yet actually see a molecule.

During that time, chemists were developing the first systematic methods of altering molecular structures, and they used the resulting changes to infer the structures of the molecules themselves—a special form of learning by doing.

The idea of atoms, of course, had been around since antiquity. In Greece circa 400 BCE Democritus argued that matter must ultimately consist of indivisible particles—as indeed atoms are, barring nuclear reactions. In Rome circa 50 BCE Lucretius argued the same case in considerable depth and suggested that dust motes that could be seen dancing in sunbeams were, in fact, driven by what is now called "Brownian motion," the effect of collisions with atoms (and for some of the motions he saw, he was right). Today, the most advanced forms of atomically precise fabrication rely on this Brownian dance to move molecules.

After classical times, centuries passed before any further progress was achieved in understanding the atomic basis of the material world. Inquiry reached a landmark in England in the early 1800s when John Dalton observed that chemical reactions combined substances in fixed proportions and explained these proportions in terms of atoms. Dalton postulated that each pure chemical compound consisted of particles— "molecules"—each containing a fixed number of one or more kinds of atoms. Reasoning from observed proportions, chemists applied this principle to infer the atomic composition of molecules, eventually deriving the chemical formula CO_2 for carbon dioxide, H_2O for water, and so on. Along another line of inquiry, the laws that describe how gases expand and contract in response to changes in pressure and temperature were explained in terms of molecular motions driven by thermal energy, the same thermal motions that cause the Brownian dance.

This kind of indirect evidence, steadily augmented with observations of the results of chemical reactions (in fact, tens, hundreds, thousands of reactions), was what led chemists to formulate and test hypotheses about the atomic structure of molecules. The systematic experiments gave rise to systematic techniques for organic synthesis, a technology of atomically precise fabrication that has changed industry, medicine, and daily life. The most impressive examples of atomically precise structures, however, came from biology. And what's more, some of these structures were functional devices that came to be known as *molecular machines*.

The concept of molecular machines dates to the mid-twentieth century and emerged out of efforts to understand how enzymes worked and how biomolecules fit together. Indeed, as early as 1890, the German chemist Hermann Emil Fischer had suggested a "lock-and-key" model for how enzymes select specific molecular substrates out of the sea of different molecules in a cell; his suggestion provided the first inkling of how complementary macromolecular shapes could enable specific parts to fit together and perform useful operations.

Since the 1950s, molecular biologists have expanded and deepened our understanding of how large molecules—including nanoscale objects made of protein—bind, move, and perform useful functions, like copying a strand of DNA in a cell's nucleus, or pulling protein fibers in a muscle to move a leg. Over the years, more and more biomolecular structures have been mapped in atomic detail, first a few in the 1950s, and today, tens of thousands, earning Nobel Prizes for James Watson, Francis Crick, and Maurice Wilkins for their discovery of the structure of DNA and for John Kendrew and Max Perutz for their use of X-ray diffraction techniques to provide the first atomically precise maps of protein structures.

This emerging knowledge of biomolecular machinery intrigued Richard Feynman. At Caltech in 1959, speaking after dinner at a meeting of the American Physical Society, Feynman discussed the physics of artificial micro- and nanoscale machinery, "inspired by the biological phenomena in which chemical forces are used in a repetitive fashion to produce all kinds of weird effects (one of which is the author)." In his talk, "There's Plenty of Room at the Bottom," Feynman proposed the idea of using machine-guided motion to assemble molecular structures with atomic precision.

Thus, in 1959, Feynman had outlined the fundamental physical principle of atomically precise manufacturing. The idea of using machines to build with atomic precision, however, then lay fallow for more than a decade and a half, while the biomolecular sciences moved forward.

By the mid-1970s, biomolecular machine engineering was already on the horizon. Scientists were beginning to write instructions coded in

DNA, founding a field they called "genetic engineering." Through this technology, scientists learned how to reprogram the molecular machinery of cells to produce new proteins—or, to speak more precisely, they had learned how to program cells to produce proteins already made by other cells.

Genetic engineering and molecular biology pioneered new fields of science and technology, but they could also be seen as opening the door to manufacturing in a new environment and on a scale that could be important for the future of humanity. I followed this field with particular attention, and by 1976 my thoughts were drawn to the question of where it might lead. (Libraries were the cause once again. As an information omnivore, I'd been casting a net into the flow of knowledge that crossed the new-journals shelves of the MIT Science Library.)

The following spring, after toying with ideas about computing with molecular devices, I found myself asking several crucial questions—not just "What could be built by programming nature's machines?" but a question a step beyond: "What could be built using the machines that nature's own machines could be programmed to build?" And then, another question a further step beyond: "What could be built using machines that could be built using *those* machines?" and so on, looking up toward the heights of a dizzying spiral of ever more capable fabrication technologies.

This upward spiral leads toward powerful manufacturing capabilities, atomically precise and yet thoroughly nonbiological, capabilities limited not by the properties of the biomolecular materials and devices that nature has evolved, but by the properties of materials and devices within bounds set only by the limits of physical law. In other words, this concept of an upward spiral suggested a path from today's technologies to advanced APM, a path based on using atomically precise fabrication technologies to build better tools for atomically precise fabrication—and a path that could begin by building with tools already at hand.

Much of the progress that has been achieved since the 1970s builds directly on biomolecular machines and materials, and this includes the rise of a field called "protein engineering." To understand the implications of progress in protein engineering, however, requires breaking away from a natural misconception.

It is tempting to think of protein molecules as watery, gelatinous stuff like meat, but this idea is highly misleading. Protein molecules are solid, nanoscale objects, much like bits of plastic, but with more diverse and intricate structures. In fact, they consist of folded polymer chains built from a kit of twenty distinct monomers that differ in terms of size, shape, and physical properties. In different combinations and sequences, these monomers can form materials as diverse as soft rubber, hard plastic, and fibers stronger than steel (spider silk, for example, is made of protein, as is the horn of a bull). But what matters most to an engineer, however, is what these nanoscale objects can *do*.

Nature shows some of the possibilities. Looking at the molecular machinery of life, we find that proteins can fit together to form motors, sensors, structural frameworks, and catalytic devices that transform molecules; protein-based devices also copy and transcribe data stored in DNA. Most important of all, machine systems built of biomolecules can serve as programmable manufacturing systems that build components for new molecular machines.

Examples like these made it clear from the start that genetic engineering offered access to tools that could make all of these things, and much more, if we mastered the arts of protein engineering.

Confronted with these facts, my thinking went something like this:

1. Nature shows that molecular machine systems can be programmed by instructions encoded in DNA to build complex, atomically precise structures, including components that fit together to form molecular machine systems.
2. Nature also shows that molecular machine systems can bind and position a wide range of reactive molecules, guiding their encounters in order to build atomically precise biomolecular structures and machine components.
3. Similar machine systems could be used to bind, position, and combine an even wider range of reactive molecules, not all found in biology, and thereby build a greater range of

atomically precise structures, including machine components that are more densely bonded and hence more robust.

4. These more robust next-generation components could be used to build robust and higher-performance production machinery, which in turn could be used to build a yet wider range of components, and from these components yet more capable production machines, and so on, extending toward a horizon far beyond biology.

In the end, to look toward this horizon means asking what physical law itself allows, and from this perspective—and viewing the landscape through the lens of systems-level engineering—I got a first glimpse of the potential power and scope of atomically precise manufacturing.

The prospects were startling. Indeed, I found them hard to believe, yet over time, studies based on exploratory design concepts, calculations, and the knowledge I gleaned from textbooks and journals persuaded me that the startling prospects were entirely realistic.

Driven by a mission to explore and share insights about the potential of these world-changing technologies, I moved to publish. First came a scientific paper in 1981,* followed five years (and three drafts) later by a book for the general public, *Engines of Creation*.

Published in September, 1986, *Engines of Creation* introduced a new concept and word into public discussion: "nanotechnology." The press formed an impression of what this might mean and ran with it.

Two months later, a leading general-audience, science-oriented magazine of the day, *OMNI*, confronted a million readers with a blazing cover headline: "NANOTECHNOLOGY: MOLECULAR MACHINES THAT MIMIC LIFE."

The article's author, Fred Hapgood, had been with the MIT Nanotechnology Study Group I had founded the year before. This enormous (and unsolicited) kickoff launched a wave of stories in newspapers and

---

* The paper, in the *Proceedings of the National Academy of Sciences,* came to be widely cited in the scientific literature as a foundation for the concepts of both protein engineering and advanced nanotechnologies based on machine-guided molecular assembly.

magazines that brought the concept to a wide audience, while the article's biological spin (MACHINES THAT MIMIC LIFE) marked a trend that grew into a problem—analogies to biology became a simplistic distorting lens though which nanomachines were mistaken for nanobugs.

Time passed, and as ideas about nanotechnology echoed and spread throughout society, they evolved and diversified to exploit a range of memetic niches. By the early 1990s, the initial, revolutionary vision of nanotechnology had launched a wave of excitement for everything "nano," and although that excitement took various forms, one form became central. As it made its way into science, the vision of nanotechnology spurred a fresh, more unified focus on nanoscale phenomena, both within science and among the general public, and it gradually grew into a surge of support for new research initiatives.

"Nanotechnology" broadened to embrace far more than nanomachines and atomically precise fabrication. It became a generic term defined primarily by size. This new, generic brand of nanotechnology (often better called "nanoscience") spanned a host of fields that worked with nanoscale structures, and it brought together researchers working with materials, surfaces, small particles, and electronic devices. They shared concepts and techniques, formed collaborations, and explored new frontiers in science and technology. The long-range vision of advanced nanotechnology excited the public, while a growing understanding of the practical importance of nanoscale phenomena stimulated the growth of both research and funding.

The story of nanotechnology, and of APM, soon became entangled in the special kinds of confusion that thrive at the borders between engineering and science. The problems stemmed from contrasts between the two that are profound yet often unrecognized.

Scientists and engineers, on the whole, are different species and have different approaches to knowledge. Scientists inquire; engineers design. Scientists study physical things, then describe them; engineers describe physical things, then build them. Scientists and engineers ask different questions and seek different answers.

In saying this, I am painting with a broad brush; a more nuanced approach recognizes that inquiry and design are often integral to a single line of research and may mesh within a single mind as it follows a single train of thought. In Chapter 8 I will draw more careful contours around questions of knowledge, practice, and culture, exploring the contrasting parts of the engine of progress that drives the modern world.

Experience shows that mistaking engineering questions for scientific questions can create a conceptual muddle. In the molecular sciences, in particular, these two modes of thought are in essence distinct, yet inextricably linked in practice and easily blurred. The costs have included missed opportunities to apply scientific knowledge to open new fields.

Scientific inquiry long ago uncovered the fundamental principles of molecular physics, and these principles enable a vast range of reliable, predictive calculations. Experimental science, however, provides knowledge—and *know-how*—beyond reach of calculation. Laboratory researchers develop what are often hard-won techniques for building atomically precise nanoscale structures, and in the course of their research the tasks of learning and making become deeply entangled. Indeed, in the beginning, it was making molecules that enabled chemists to discover atoms and bonds, long before quantum mechanics provided an explanation.

Thus, when my work led me into the land of the molecular sciences, I found a culture in which questions of inquiry and design were often confused, a culture in which most researchers scarcely recognized the concepts and methods of systems-level engineering. Nonetheless, I found that abstract engineering concepts had direct applications.

Consider, for example, the pattern of thought then prevalent in protein science. A scientific problem—given a monomer sequence, predict how a protein will fold—had been confused with an engineering problem—given a desired fold, design a monomer sequence that will produce it. At the time, fold prediction was an unsolved problem (and remains only partially solved today), and researchers had implicitly assumed that successful fold prediction must precede fold design.

My 1981 paper, however, explained why design and prediction were fundamentally different problems and why design should be less challenging. Fold design was soon dubbed "the inverse folding problem," and this deep, elementary idea launched the field of protein engineering.

Protein engineering, however, remained embedded in science. Speaking at a conference a decade later, I asked for a show of hands: "How many of you consider yourselves to be scientists?" and about one hundred hands went up. When I then asked, "How many of you consider yourselves to be engineers?" the total was no more than three. And this at a conference convened with the title topic, "Protein Engineering."

Fields differ, of course. If I had asked the same questions when speaking to an audience of experimental physicists, I suspect that many would have raised their hands twice, and likewise in talks I have given to space scientists tasked (for example) with sending instrument systems to Mars.

As a rule, however, one finds that engineers and scientists have contrasting cognitive habits, intellectual values, and cultures; the contrast is particularly sharp where science centers, not on complex systems like spacecraft or particle accelerators, but instead on laboratory experiments using equipment appropriate to the molecular sciences (think of beakers, pipettes, commercially available infrared spectrographs, and the like). In the molecular sciences, most researchers have had no reason to learn the arts of systems-level engineering design.

Thus, the modes of thought that were best suited to the needs of molecular research were ill-suited to the task of grasping or judging abstract engineering analysis, while the researchers who could easily understand the scientific basis for atomically precise manufacturing were liable to slip into confusion about the concept itself, and misjudge it. Then came a push, a pull, and a slide into a conceptual pit.

THE ENGINES OF CONFLICT

The vision presented in *Engines of Creation* unleashed forces that soon came into conflict. The first, of course, was the force of vision itself, which spurred studies that deepened our understanding of the prospects. As part of this effort, my own attention had turned toward preparing my doctoral dissertation, and from this, *Nanosystems*.

The many-sided force of popularization, however, had a head start of six years. As they echoed through the press and Internet newsgroups, the concepts presented in *Engines* devolved into vivid, simplistic stories and images and the science and engineering lost ground to fiction that shaded off into ideas that amounted to magic. Utopias and scare stories took form and gained strength. From a distance, the concepts in *Engines* were perceived through layers of distortion and science-free fiction.

Excitement and popularization presented a risk because when presented in a brief or distorted summary, APM-level technologies seemed like hype gone wild, and when wrapped in popular enthusiasm, the concepts seemed to be no more than another delusion of crowds. Ironically, what makes these technologies important—the radical scope of their implications—makes them hard to credit for reasons that are correct at least 99 percent of the time. Heuristics, however, sometimes go wrong, and this is an instance.

In the early days, however, from where I stood, it seemed that enthusiasm was primarily a positive force—that it would (as it did) channel support toward scientific progress, and that scientists, in turn, would surely help to channel enthusiasm toward reality and gradually push nonsense into the background.

Indeed, this happened to some extent as reality-based thinking continued to advance. Students read *Engines* and turned their careers toward nanotechnology; researchers at Caltech and elsewhere applied computational methods to study advanced AP machinery; I spoke at scientific conferences, corporate meetings, the White House Office of Science and Technology, the Pentagon, NSA, the Congressional Office of Technology Assessment, and at a Senate hearing convened by then Senator Al Gore.

By the end of the decade, researchers in a host of fields (but weighted toward materials science) had gathered under the banner of nanotechnology, both promoting and stretching the vision attached to the word. By the late 1990s, support for the resulting, greatly broadened kind of nanotechnology had reached the threshold of launching a federal program, making ownership of "nanotechnology" a billion-dollar prize.

Near-term research and longer-term objectives were entirely compatible, or should have been, yet as events unfolded, a conflict emerged, feeding on clashes between popular visions and near-term scientific realities. This conflict polarized, taking on an us-vs.-them, and science-vs.-fantasy tone, while the distinction between fantasy and genuine prospects was increasingly lost in the noise. A turning point came when the new federal program's promoters secured funding to develop atomically precise fabrication: They turned against the vision they had sold to Congress, redefined their mission, and launched a strange and confused war of ideas that still echoes today. I will pick up this story in Chapter 13.

The conflict had a particularly perverse effect. It severed "nanotechnology," as widely perceived, from the concept of atomic precision and its natural roots in the molecular sciences. I had outlined a path toward APM that led forward from existing capabilities for atomically precise fabrication, yet strong, continuing progress on that very path somehow slipped from attention. This has set up the world for a surprise.

Consider how far we have already come. In 1986 neither protein engineering nor structural DNA technologies had yet been demonstrated, no one had yet used a machine to move individual atoms, and the largest complex, artificial atomically precise structures were no more than a few hundred atoms in scale. Since then, atomically precise fabrication technologies have made great progress on multiple fronts:

- Researchers now routinely use scanning probe instruments to image and place individual atoms and to maneuver and bond individual molecules. This level of control has demonstrated the principle of mechanically directed atomically precise fabrication.
- Organic chemists have built steadily larger and more complex structures along with motors and other machines; their techniques now provide a rich toolkit for building molecular systems, while inorganic chemists and materials scientists have expanded a complementary toolkit of nanoscale structures.
- Protein engineering has flourished, supported by computeraided design software, and now enables the routine design of intricate, atomically precise nanoscale objects, including structural components and functional devices.

- Structural DNA nanotechnology has emerged and now enables rapid and systematic fabrication of addressable, atomically precise frameworks on a scale of hundreds of nanometers and millions of atoms.
- Quantum methods in chemistry have advanced together with the power of computers and algorithms, providing powerful, physics-based tools for scientific modeling and molecular engineering.
- Molecular mechanics methods in chemistry can now describe the structure and dynamics of molecules on scales that reach millions of atoms, a range that can enable the design and development of complex, atomically precise systems.

The current state of the art is more than enough to support a drive for next-generation molecular systems on the road to atomically precise manufacturing. Indeed, I am persuaded that advances in recent years now provide a platform that can support rapid progress. The greatest challenge today is to put the pieces together—not only components and computational tools, but also engineering concepts and the research teams that can bring them into physical reality.

INTERLUDE: PROSPECTS AND CHALLENGES

Stepping back for a moment, let us ask a question: "Where do we stand today as we consider APM-linked technological choices that could change the shape of our future?" The issues reach far beyond laboratories, politics, and molecules.

In brief, the APM-based production revolution promises to transform the material basis of human life with far-reaching consequences that include both new solutions and new problems of global scope.

Consider the challenge of resource scarcity (minerals, petroleum, water) and the challenge of environmental problems that range from toxic metal emissions to global climate change. These are physical problems that have potential physical solutions. Through a chain of physical and economic links, the APM-based production revolution can transform global problems by slashing resource consumption and toxic emissions and by providing the infrastructure for low-cost solar energy and a carbon-neutral economy (and even more remarkably, by providing affordable means for removing carbon already released into the atmosphere).

These physical capacities could solve critical problems and enable us to live more lightly on Earth, while radically raising the material standard of living worldwide. These solutions bring problems, however. In particular, rapid deployment of this range of capabilities would lead to deep, pervasive disruptions in the global economy, beginning with mining, manufacturing, and trade, and spreading outward from there.

How would APM have these far-reaching effects? In manufacturing, APM-based technologies can produce better products at far lower cost than today, out-competing existing industries. As for mining, APM naturally consumes and produces a different mix of materials (no need for the iron and chromium in stainless steel, no need for lead and tin in solder) and as it happens, the most useful elements—including carbon, nitrogen, oxygen, and silicon—are not at all scarce. On grounds of performance and cost, even common structural materials will be subject to widespread displacement, and with them, most mines. (Chapter 11 explores questions of APM-based product performance, cost, and resource requirements in greater depth.)

Today, trade builds global supply chains that lead from mines and wells to smelters and refineries, to materials processing plants, to networks of factories that shape and assemble components to make final products. As we will see, with APM-based technologies it would be natural for these long, specialized supply chains to collapse to a few steps of local production, progressing from common materials to simple chemical feedstocks; from simple feedstocks to generic, microscale building blocks; and then from generic components to an endless range of products, much as printers can be used to arrange generic pixels to form an endless range of images.

Long, specialized supply chains drive the physical trade that today joins the world into a global economy, and collapsing supply chains would cause that trade to decline. One can easily imagine disruptions in trade that would affect the livelihood of half the planet or more. And one can easily imagine a level of suffering and scarcity in the midst of potential abundance.

I think that this prospect points to a need for exploring policies for managing what could be a catastrophic success. In other words, it calls for a conversation that considers prospects for our world as the physical potential of APM-level technologies crosses the threshold into physical reality, a conversation that was interrupted more than a decade ago and must now be renewed.

Implications for the military sphere, in particular, demand careful consideration because easy, unconsidered policies would bring great and needless risks. Here the nature of potential products (and of the potential dynamics of their development, production, deployment, and use) will have profound implications that call for fresh thinking. The economic implications of an APM transition likewise call for a reassessment of national interests as deep and broad as the prospective changes in the material economy.

Enough can be understood today to reframe global problems and raise new concerns. In a world on the path to profound transformations, our situation calls for asking unusual questions about our prospects and how we might best respond—new questions of how to avoid needless risks, resolve difficult global problems, grasp unexpected opportunities, and manage disruptive change.

In short, we need to begin to broaden the agenda for conversations about the future—not to change widespread premises in an instant, but to begin to assess the prospects for APM-level technologies and the questions they raise regarding challenges and prudent near-term choices.

———

MY AIM, HOWEVER, is not to overturn anyone's worldview. Prospects for radical abundance deliver a banquet of almost indigestible truths— or so the prospects seem to me, even now. Digesting and integrating new information will take time and the contributions of many minds.

My aim is, therefore, more modest: to outline facts about what is truly possible, to discuss where technology may lead in the coming years, and to consider some critically important questions that have not yet been asked. In light of the prospects ahead, I think that it's time to begin a new conversation about our future, a conversation that begins to explore the prospects for radical abundance.

PART 2

THE REVOLUTION IN CONTEXT

CHAPTER 4

Three Revolutions, and a Fourth

THROUGH NEW TECHNOLOGIES, human history has repeatedly changed directions, and with unimaginable consequences. The Agricultural Revolution of the Neolithic era marked the advent of a way of life that enabled dense, settled populations to feed themselves and opened the way for the rise of civilization. The Industrial Revolution launched an explosive wave of physical products that brought with it the modern world. The Information Revolution, still unfolding today, has rewoven the fabric of knowledge, commerce, and society, setting the stage for a future whose outlines are still beyond knowing.

The approaching APM Revolution will provide the driving force for a fourth revolution, and like the preceding revolutions, it will transform daily life, labor, and the structure of society on Earth.

Past revolutions offer lessons that can help us grasp the nature of the revolution ahead. Each has had pervasive impacts on the human world, bringing profound and far-reaching change, and the very nature of their technologies can help us understand how the APM revolution compares with those of the past.

The Agricultural Revolution was based on molecular machinery (though not machines made by design); the Industrial Revolution was based on machines made by design (yet far from molecular); and the Information Revolution was based on digital, nanoscale devices (but devices that process only information, not matter). In a sense, the APM Revolution draws from all three of these. It employs artificial, molecular, nanoscale machinery that operates on digital principles (but this time, in order to process matter). It should be no surprise, then, that each revolution holds lessons for understanding the APM Revolution and why its advent will mark a divide in human history.

As Winston Churchill once said, "The farther backward you can look, the farther forward you are likely to see." Here, we can begin with the Neolithic era.

THE FIRST AGRICULTURAL REVOLUTION

The first agricultural revolution—*the* Agricultural Revolution—defines the dawn of the Neolithic era, more than ten thousand years in our past.

The Agricultural Revolution gave human beings a new way to exploit the productive capabilities of the nanoscale machines found in living organisms, the molecular metabolic machinery that makes complex structures out of nothing more than water, soil, air, and sunlight. Agriculture multiplied the food yield of land by a hundredfold, yet the agricultural way of life also marked the beginning of a cascade of change. For example, the Agricultural Revolution dates from prehistory for a simple reason: Writing came later, and made written histories possible.

Our knowledge of the time before written history grows richer every day. Archeologists and geneticists read unwritten records of earlier times in sequences of artifacts layered in soil and the patterns inscribed in our genes. Neolithic artifacts show the emergence of a new tool technology, with stone ground into shapes suited for felling trees, shaping wood, and grinding grain—new, robust tools unlike the keen but fragile chipped-flint arrowheads that had been used by hunters for untold generations.

Alongside these Neolithic tools archaeologists find other markers of change—traces of grain from the fields farmers tended and pits left by long-rotted timber posts that had supported the roofs of forgotten settlements. From these artifacts and clues, archaeologists have dated the transformation. It began over ten thousand years ago, at the end of the Pleistocene ice age. In a warming, hospitable climate, people gradually shifted from gathering grains to gathering, then saving, and *planting,* some of the best seed from the previous year's harvest. Indeed, the archeological record displays the evolution of larger, more modern forms of grain, showing the effects of artificial, unnatural selection. In Mesopotamia, the cycle of harvest and planting soon gave rise to what we know as einkorn wheat, barley, lentils, and chickpeas.

Other centers of agriculture arose independently in China, New Guinea, Africa, and the Americas, but it was Mesopotamian agriculture that ignited the conquest of Europe. Markers in soil and genes show the spread of a wave of farmers northward and westward, to the ends of the continent. As these early technologists spread, they settled where the land was most fertile and pushed the hunter-gatherer way of life to the margins.

Why did agriculture displace hunter-gatherer life? Not because it was better for people, families, or tribes. Leisure time decreased, because tilling and tending the Earth called for steady, hard labor to build food stores, yet when groups neared their limits to growth and production encountered a bad harvest year, a springtime of starvation could follow regardless of effort. Hunting and gathering, by contrast, was an on-again, off-again business, with results more subject to luck and less subject to labor; a season of abundance might follow or not, regardless of the amount of labor in the season before. Among hunter-gatherer societies, limits had a different form, subject to patterns of scarcity, abundance, diet, and leisure that shaped different ways of life, and people.

The agricultural way of life eroded both leisure and health. Stature decreased and teeth decayed as human beings' previously eclectic diet became reliant on the few crops they could produce in bulk.

What *did* increase were population density and the strength of ties to plots of land. With an ever-growing numerical advantage, even the less able-bodied farmers could enter new lands and extend the domain of agriculture one plot at a time.

Hunter-gatherers could fight back, of course, but farmers had more of the primary resource for defense and war: people. Thus, the spread of farming had nothing to do with overall human well-being. The gradual conquest of continents by the Agricultural Revolution was an unintended consequence of a new way to get food. More consequences followed.

With their tools and hard labor, farmers stored grain as a buffer against starvation, bringing another unintended wave of social development: their food stores enabled new ways of life using weapons as tools and harvesting grain from the farmers themselves, almost as if they were crops. Like plants, farmers were rooted to the land—a resource worth raiding and plundering or defending and taxing. Both the pull of trade and the need for defense gave rise to cities, empires, and civilizations that evolved through growth, collapse, and succession, growing in scale and societal complexity. And thus with cities and trade came specialized tasks, tools, skills, and knowledge, steps on a path of rising technology.

Technology arose from wealthy cities; wealthy cities arose from farmers rooted to the soil; the farmers' way of life, in turn, arose from gathering, saving, and planting seeds—all this, and the conquest of continents, followed from the requirements of agricultural ways of producing food and from the need to store it.

Thus, the ultimate driver of change at the dawn of civilization was the physical nature of a technology—agricultural production—and its demand for land, soil, water, and labor to harness the energy of sunlight for the production of food from plots of land tended from season to season. And the need for this set of resources, in turn, stemmed from the physical nature of the machinery of metabolism, the conditions required for particular kinds of nanoscale machines to process molecules with atomic precision.

THE INDUSTRIAL REVOLUTION

Millennia after the Agricultural Revolution, and in just a moment by the measure of history, the Industrial Revolution gave rise to the modern era. Three lifetimes laid end to end could span the time since the revolution began.

Advances since the takeoff of industrial society (which took hold in Britain around 1800) have multiplied the developed world's productive capacity by a factor of one hundred or more, if one can compare products as different as wagons and aircraft, or books and computers. Indeed, the vast proliferation of different kinds of products has been more striking than the growth of quantity—industry now produces not merely more cloth and carriages, but nylon fabrics and horseless carriages, and exotic products like aircraft, telephones, software, and nuclear bombs.

The Industrial Revolution embraced myriad machines and manufacturing methods, new methods of engineering complex systems, and new modes of management and governance to orchestrate growing complexity. If one looks for the core of modern technological capabilities—the basis of a world able to make things like aircraft and smartphones—one finds that those familiar products are made with the aid of machines built by means of machines that were built by means of yet other machines—tools used to build tools in an unbroken chain that leads first to distant factories, and then deep into time.

Tracing this chain backward through history, one finds tools forged by blacksmiths. In the distance, we can hear the echoes of craftsmen using hammers and tongs to forge iron tools that their apprentices mastered and bettered, advancing through stages and centuries to make high-precision tools and machines, and then mechanized tools and yet better machine tools in a process that continues today.

The core of progress in production—and hence in technology—is like the trunk of a tree with roots deep in history. The products that we see every day are like the leaves and fruit of this tree, while behind the scenes the factories that manufacture these products are like branches that grow from the trunk, from a technological infrastructure of precision machines that collectively make all the precision components that are required to build machines of equal precision, or better.

In today's world, the trunk of this technology tree is all but invisible. The core production machinery doesn't even make the components you'd see if you were to pop open the hood of car, or crack open the case of a phone, or study a chip under an electron microscope. Nonetheless, this central core supports all of the rest, producing the machines used to make the machines that make what you see, and beneath a tangle of intertwining links, core production technologies provide a platform for industrial civilization itself, a platform that continues to rise.

The Industrial Revolution that grew with this core brought a further revolution in agriculture. Just as hunter-gatherer bands gave way to agricultural villages, small farming communities have yielded to industrial agriculture. Even where towns still continue as farming communities, their peoples' concerns, tools, and labor have been transformed beyond the comprehension of a Neolithic villager pulling weeds from a garden by hand, or of a peasant plowing a field with an ox.

The Industrial Revolution had its most profound effect on human life by enabling food production to outpace population growth. Before, civilizations had been stuck in the age-old Malthusian trap, where expanding capacities increased population, but not prosperity. Despite short-term fluctuations (the Black Death brought a respite), populations always collided with their limits to growth, stunting their potential development. Escaping the Malthusian trap opened new horizons.

The spread of mechanization served agriculture directly as it advanced from horse-drawn reapers that cut wheat in the mid-nineteenth century to tractor-driven machines and then combines in the twentieth century, reaping, threshing, and winnowing, leaving few if any human footprints in the fields. In a further, crucial development, industrial products began to nourish the plants that fed the human race. Shortly before World War I, Fritz Haber developed a process that made ammonia from atmospheric nitrogen, a step in the production of explosives for war, but also an invention that provided the crucial ingredient for nitrogen fertilizers, which have since been used to multiply crop yields on every continent. Indeed, most of the nitrogen in your body very likely arrived in your food through the Haber process.

In countries where both famine and actual farming are now largely forgotten, the industrial mode of production has transformed almost every aspect of life: how we work, what we eat, where we travel, and who we can see and hear—not only those alive and nearby, but also faces and voices speaking from distant continents and distant years.

The explosion of industry brought with it an explosion of specialties and the knowledge these specialties embody. In the early days of the Agricultural Revolution there were no more than a handful of trades making practical things like axes and baskets from stone, wood, and reeds, then millennia later, trades that made things of bronze, iron, and steel. By the dawn of the Industrial Revolution, complexity had increased dramatically, with craftsmen working in iron and steel to make products that ranged from anvils to clockworks and (famously) manufacturing pins, a task in which, as Adam Smith noted, specialized tools and division of labor could multiply productivity by a factor of hundreds, or thousands.

In yet another unintended consequence, the rise of new specialties in manufacturing spurred the growth of science not only by providing new scientific tools, but also by creating an insatiable demand for science-based practical knowledge.

Metalworking raised questions that spawned the science of metallurgy, and early metallurgy, in turn, spawned seemingly endless specialties (the study of alloys, thermal processing, and dispersion hardening; corrosion and oxide film formation; the role of crystal dislocations in plastic deformation and metal fatigue . . .) and these specialties continue to proliferate from year to year as materials, instruments, and applications multiply. The venerable ranks of steel hammers and nails, rivets and I-beams now share the field with materials as different as thin metal films and single-crystal superalloy castings, which in turn serve purposes as diverse as nanoscale wiring for silicon chips and high-stress, high-temperature turbine blades for jet aircraft engines. Today, metallurgists draw on a host of techniques and instruments, studying materials through mechanical stress, chemical analysis, electron microscopy, and X-ray diffraction, techniques that can reveal the properties and structure of materials, from the scale of an airplane wing down to patterns of atoms.

Like scientists in thousands of other fields and specialties, metallurgists draw on the abilities of engineers, instrument manufacturers, and journal publishers, to say nothing of those who enable them to meet and share knowledge (hotel-builders and managers; aircraft designers, rivet-makers, and pilots) and also, of course, a handful of people they actually know—colleagues from across the hallway or from the far ends of the Earth.

The modern division of knowledge complements the classical division of labor. A blacksmith in a long-ago village shared a body of traditional knowledge with other blacksmiths who lived within range of travel by horse or on foot, but modern scientific meetings can draw researchers from anywhere in the world, each one bringing unique expertise. Specialization has, in a sense, neared its ultimate limit—each expert having different knowledge and tasks—an advantage for the growth of deep, detailed knowledge, yet a barrier to efforts to understand how the parts fit together. (And turning to literal, physical parts, AP structures made by researchers in the molecular sciences are today's most important examples of parts not yet made to fit together.)

Systematic research and development itself was a product of industry, because industrial corporations established the first R&D laboratories. Some of the earliest laboratories appeared in Germany during the late 1800s. The very first, established by Krupp, studied the metallurgy of alloyed steels; other laboratories developed methods for atomically precise fabrication of sub-nanometer structures (which is to say, methods for organic chemical synthesis). In the United States, Thomas Edison's "invention factory" at Menlo Park provided another early model for organized research, and during the early twentieth century industrial R&D labs proliferated (established, for example, by General Electric, Westinghouse, Bell Telephone, and DuPont). During World War II, the US federal government greatly expanded other dimensions of research support, establishing a series of National Laboratories, and in 1950, the National Science Foundation. Other industries and governments soon followed or led, and global research and development expenditures now exceed $1 trillion per year.

The interplay of research fields at different levels proved intensely productive. Quantum physics, for example, emerged from scientific research that aimed at no practical purpose, yet the resulting knowledge has transformed the foundations of technology. Applied to the study of semiconductors, for example, quantum physics gave Bell Labs researchers the insights that led them to invent the world's first transistor—a crucial step on the way to the nanoscale devices that power today's ongoing revolution in digital information technologies.

Ample food and increased production had opened doors that led to yet more advances. Agriculture enabled civilization; industry rebuilt it.

THE INFORMATION REVOLUTION

The Agricultural Revolution dates from prehistory, while the Industrial Revolution began to gain steam not long before yesterday. A single lifetime, however, is enough to span the whole of the Information Revolution, from its barest beginnings with racks of vacuum tubes to its role in today's unfathomable acceleration of global change.

I find it hard to imagine life without access to Internet search, even though I lived such a life until I reached middle age. Wikipedia in over forty languages, a billion Facebook users, global access to free online courses from MIT—all these and more are among the landmarks today, along with destructive, but not necessarily unhealthy, pressures on traditional media and retail distributors, the healthy pressures of online reputation systems, the drift away from reality within information bubbles, and a flood of both insightful and unhinged online discussion.

Today, most readers will have experienced a massive change in the texture of the world's conversation, both for better and worse, with the rise of media that carry rumors and news between New York and New Zealand more quickly than the filtered streams of news that had once flowed between uptown and downtown Manhattan.

The Information Revolution is intertwined with the previous revolutions and has also transformed science, engineering, and the production technologies that support the physical aspects of change. The Information Revolution itself both draws on these advances and enables them, driving forward the line of progress in science, engineering, and semiconductor fabrication that has enabled invisibly small devices to replace glowing vacuum tubes the size of your thumb.

The first stored-program computers were built in the United Kingdom in 1948 and 1949 at the Universities of Manchester and Cambridge.

These early machines used vacuum tubes to build digital systems; the first fully transistorized computer went operational in 1955, the year I was born.

Compared to the machine on the table in front of me now, the first practical computing machines could store less than one-millionth as much data, while their speeds lagged by a factor of one hundred thousand. They often consumed more power than a house, weighed more than a ton, and cost more than a million present-day dollars.

From this low baseline capacity, exponential improvements in performance have followed, and for half a century now the leading technology has been in essence the same—transistorized circuits on the surface of silicon chips—and the pace of progress has been set by device miniaturization. This exponential progress, which held steady for decades, is dubbed "Moore's Law," which states that every two years ongoing miniaturization doubles the number of transistors per chip. At this pace, two years gives a factor of two; ten years, a factor of one thousand; twenty years, a factor of one million. In 1971, Intel (co-founded by Gordon Moore, of Moore's Law fame) shipped the first machine that compressed all the components of a computer processor onto a single chip, the 4004, and it outperformed machines of the kind that decades before had weighed tons.

Agricultural production sustained agriculture in a physical sense, and industrial production provided the physical basis for industry. The Information Revolution, however, is in a sense derivative, a product of progress in the industries of semiconductor manufacturing. Its most fundamental advances are not in the realm of *information,* but instead in the realm of *physical production,* a tightly coordinated set of advances in the light sources, optics, masks, and steppers that shape nanoscale patterns of plastic on chips, along with advances in ion implantation, vapor deposition, and etching processes that transform those patterns into metals, insulators, and silicon transistors.

On its physical side, the Information Revolution has been an outgrowth of conventional industry with a conventional economic structure: materials produced and refined, intricate supply chains, and products that are costly to produce and costly to purchase. On its information side, however, the Information Revolution provides humanity's first example of radical abundance and how it can step beyond the usual economic framework. Standard accounting, for example, grossly mismeasures the value that the Information Revolution has delivered to the human race.

Investors in the Internet bubble of the late 1990s perceived correctly, I think, the great value of Internet services. They erred, however, in expecting much of this value to materialize in the money economy of income, spending, and corporate revenue. Indeed, conventional accounting is blind to most of the value that people receive.

How much are Wikipedia's services worth? Many billions of dollars per year, by any reasonable measure. Yet how much revenue does Wikipedia get, as an organization, and what is its dollar contribution to a nation's GDP? Both are zero, or nearly enough. Likewise for all the free content found on the Web and the entire world of open source data and software. These compete with costly products and deliver similar value, yet downloading and using free, open-source products leaves no footprint in corporate ledgers or aggregate financial statistics. If the use of free products reduces GDP per capita, then so much the worse for the idea that GDP measures value.

A simple, radical fact underlies this economic anomaly. The marginal cost of copying and delivering information goods is often so close to zero that even for-profit businesses offer these goods in return for no pay other than to bring their names and achievements to the attention of peers and consumers, and for the personal satisfaction that comes from solving problems that matter to people.

The concept of "production" has changed. In the material world, manufacturers incur a cost for designing a product (where a design is a pattern encoded in data), but also a cost for producing and shipping each physical unit. In the digital information world, by contrast, what we call "production" yields a pattern encoded in data—in effect, a design— while true "production," in a manufacturing sense, is the process of copying data from one place to another at a cost not even worth charging.

THE APM REVOLUTION

What can we say about the coming APM Revolution, seen from the perspective of the three previous revolutions? A great deal, as it turns out. Both parallels and contrasts offer insights.

The biological machinery of the Agricultural Revolution, the digital nanoelectronics of the Information Revolution, and the innermost mechanisms of APM systems all show ways in which nanoscale devices can be harnessed within macroscale systems to deliver useful results. In other words, three of these four revolutions rely on nanoscale devices, the Industrial Revolution being the only exception.

However, although the Agricultural Revolution employs atomically-precise nanoscale devices, these are products neither of human action nor human design. Rather, these devices consist of soft structures that perform a Brownian dance, twisting and tumbling in the disorderly environments found inside cells. In common with APM, these biological systems process molecules with atomic precision, yet considered as systems, they have little in common with the orderly arrays of machines in a nanoscale factory.

Like APM systems (but unlike cells), the machines of the Information Revolution employ nanoscale devices linked in fixed, orderly patterns, and these devices are made of rigid materials, much like those used to build conventional machinery. Nanoscale electronic devices, however, are far from being atomically precise and, of course, they process only information.

Where does APM stand, seen from the three-fold perspective of the earlier revolutions? It employs nanoscale devices that work with atomic precision, but made of rigid materials and linked in fixed, orderly patterns more like circuits on chips than like molecules diffusing in cells. Indeed, because manufacturing tasks require patterns of motion that have no dependence on scale, the natural organization of machinery inside an APM system parallels what one would find today in a factory.

In contrast to industrial systems, however, APM-based production embodies the essence of the digital principle that underlies computation. Just as analog electronics can never reproduce a signal exactly, industrial processes can never reproduce a form exactly. And just as digital electronics can produce exact patterns of bits, so APM mechanisms can produce exact patterns of atoms, which is to say, extraordinarily large yet precise molecular structures. In this respect, APM amounts to a digital revolution transposed into the material world, because both are based on the principle of discrete, noise-tolerant operations, a result achieved by meeting well-chosen engineering constraints that stem from physical law.

And like the AP machinery of biological metabolism—and in contrast to industry—APM has no need for a globe-spanning infrastructure. An APM supply chain can be nearly as short as the path from sunlight and soil, through grass and hay, to products like milk and wool—a long metabolic supply chain, when traced in molecular detail, yet just a few links as seen by a herdsman. APM-based production can likewise rely chiefly on common, local materials, and on solar energy (through low-cost production of photovoltaic cells), with no net CO_2 emissions.

This contrasts sharply with the globalized consumer economy of the current era. Modern industrial supply chains branch out though factories and processing plants that draw on labor and scarce raw materials from distant lands. With the technology of AP fabrication, better products could be made without these inputs.

The role of human labor also differs across the four revolutions. Labor in traditional agriculture could be long and hard, but seldom required a new idea. While everyday labor in industry could be just as routine as a farmer's, the dynamics of competition and ongoing technological advances make novelty increasingly crucial; in the information world, of course, the balance has shifted further. For APM-based production, like information technologies, the crucial labor consists of creative work. In both instances, an innovation, once reduced to a digital description, can be reproduced endlessly, anywhere, by generic and fully automatic processes. In the information world, a program or image need only be copied, with negligible labor or resource cost, and for APM-based production, the situation will be nearly the same. Small amounts of labor and inexpensive resources will suffice to produce the latest products, not quite the same as in the information world, but close enough to suggest strong economic parallels.

Agriculture, digital systems, and APM have this in common: In the
normal course of events, there's no need for a person to stick a hand into
the guts of the production system—no manual work to be done in photo-
synthesis, in computation, or in APM processing. If plants and animals
were more cooperative (and less vulnerable to diseases, weeds, pests, and
predation) they'd join digital systems and APM as minimal-labor production systems.

Industrial production is already trending this way. For a century or
more, industry absorbed an increasing portion of total labor, but in recent years, that trend has reversed. Many of today's large-scale industrial
systems rarely require the touch of human hands and the extent of automation is growing. Indeed, the ultra-clean guts of semiconductor processing systems—the parts that work directly with wafers—would be
rendered useless or even destroyed by contact with human skin.

I recall touring a Korean steel mill, POSCO's Pohang plant, which
produces half a ton of steel per second. From a catwalk high above a long
and wide hall, I surveyed an array of massive machines hard at work
rolling slabs of steel. I began to look for workers, and eventually, looking
toward the far end of the hall, I caught sight of a lone man climbing stairs.

The very nature of APM-based production (highly productive capital, negligible labor, and frugal use of inexpensive materials) has the
capacity to sharply reduce the cost of products to a level almost unrelated to their cost today. Costs, however, are only one aspect of radical
abundance. The range of potential APM products includes not only
radical extensions of the current range of industrial products, but also
products that, while much like those familiar today, are made of better
materials and components that sometimes enable startlingly better performance. Through replacing low-cost steel structures with lighter,
stronger materials, the cost of typical structures could fall by a factor
of ten; through replacing and upgrading portable electronics, costs could
fall by a factor of a thousand even as the computing power delivered
grows by a factor of more than a million.

By historical standards, the prospects for APM-based production
amount to another great revolution, another divide in human history,
bringing challenges and opportunities of immeasurable scale. Agricultural

production fed more people, yet shifted their lives into patterns of hard labor and scarcity, while setting processes into motion that spawned civilizations, war, and technological progress. Industrial production provided more of everything—food, shelter, transportation, household goods—and brought an era of economic and military competition that forged today's uneasy globalized economy and shifting military balance.

The Industrial Revolution enabled production to outpace the growth of population, enabling humanity to surpass age-old limits, to escape the Malthusian trap, and to flourish as never before. Today, however, concern is rising that this revolution has reached the end of its run, approaching limits to its exponential growth. APM-based production, however, can provide an alternative, lifting the world far above the Malthusian trap, with open horizons that could, with care, stretch deep into the future of humanity.

The pace of innovation continues to increase, and the Information Revolution holds a hint of what may lie ahead. Taken together, the parallels between APM-based production and digital information systems suggest that change in an APM era could be swift indeed—not stretched out over millennia, like the spread of agriculture, nor over centuries, like the rise of industry, nor even over decades, like the spread of the Internet's physical infrastructure. The prospect this time is a revolution without a manufacturing bottleneck, with production methods akin to sharing a video file. In other words, APM holds the potential for a physical revolution that, if unconstrained, could unfold at the speed of new digital media.

But before we can come to grips with this revolution in a concrete sense, we must first gain a grasp of how the nanoscale world looks and feels—or how it would, if our senses and physics allowed. It is time to learn to think small—ten million times smaller, to be precise.

Table 1 Four Revolutions Compared

Revolution	Its Basic Nature	Its Human Impacts
Agricultural Revolution	Provided new means of producing food and materials by exploiting the productive capabilities of the molecular nanosystems found in living organisms.	Led to stable settlement and investments in the land itself, multiplied population densities by factors as great as one hundred or more. Made possible the development of cities and civilization.
Industrial Revolution	Provided new means of producing material objects by exploiting the productive capabilities of artificial mechanical systems on a human scale.	Multiplied productive capacity by a net factor now on the order of one hundred during a period of two hundred years. Made possible new products, new ways of living, and ongoing changes in the structure of civilization.
Information Revolution	Provided ways to process information using high-frequency nanoscale devices to process and deliver patterns of bits.	Multiplied information processing capacity on an exponential curve, to date by a factor of more than a billion. Made possible new information products and new ways to organize the material and human world along with ongoing changes in the structure of civilization.
APM Revolution	Will provide new means of producing material objects: Like the Agricultural Revolution, it will use atomically precise productive machinery. Like the Industrial Revolution, it will use artificial mechanical systems to make things. Like the Information Revolution, it will use high-frequency nanoscale devices to process and deliver patterns (but of atoms rather than bits).	Like each of the prior revolutions, it will lead to deep changes in products, productivity, means of production, and human society. Able to multiply the productivity of agriculture, manufacturing, and computation by factors of ten to one million. Can radically extend the scope and scale of production, transforming the material basis of civilization and reducing its impacts on climate and the Earth as a whole.

CHAPTER 5

The Look and Feel of the Nanoscale World

IT'S HARD TO UNDERSTAND how molecules interact when they're illustrated in so many different ways, sometimes as blobs or clouds of dots, sometimes as ball-and-stick models, sometimes as skeletal, structural diagrams, and sometimes as images in the popular press jumbled together with pictures of microbes and machines ten thousand times larger. Without context, these different pictures can easily lead to confusion. To get a clear view of the nature of APM systems, it will help to start with a more coherent view of the nanoscale world.

In the molecular sciences, researchers often describe molecules in familiar, mechanical terms, as small moving objects with size, shape, and stiffness. Although some kinds of molecules have unusual kinds of flexibility and exhibit unfamiliar behavior, a selected, particular range of molecular objects can be understood remarkably easily. These are the solid, rigid molecular objects of the kind best suited to building high-performance nanoscale machines, and machines made of parts like these turn out to be surprisingly similar to the larger machines of everyday use. Indeed, the machinery of atomically precise manufacturing can be visualized in a way that is both qualitatively intuitive and quantitatively realistic.

A well-chosen, magnified view of nanoscale objects and mechanical devices can provide this perspective. To keep exotic concepts well-grounded, it's important to understand how much of what matters most can be understood with little more than common sense and a few key facts.

APM, for example, deals with rigid molecular structures, and even state-of-the-art computational models in the molecular sciences describe structures like these in mechanical terms, as the material objects they are. Indeed, they can be grasped by the mind using the neural mechanisms that evolved long ago to cope with the material world.

In the brain, MRI imaging shows that visualization and visual experience overlap at a neurological level, engaging the same neural mechanisms. Almost a century ago, electromyography showed that imagined actions are reflected in subtle responses by the muscles that these actions would move; this may be why tennis players find that imagining practice improves their game. To understand the approaching revolution in atomically precise nanotechnology, we can use this human capability to activate the parts of our minds that are adapted to grasping tools with our hands.

When scientists play with models of molecular structures through a computer interface, they gain a visual, even kinesthetic, understanding of how these objects behave. Experimental methods and computational chemistry can provide not only quantitative knowledge, but also an accurate, intuitive, physical sense of molecules as objects. This intuitive grasp can complement and refine expectations founded on calculations and scaling laws.*

Molecules have size, shape, mass, momentum, strength, stiffness, and so on, and larger molecular structures become more and more like familiar objects.** Indeed, some large molecules are familiar objects; a

---

* The quadrupling of area when linear size doubles is a scaling law; the proportional scaling of strength with cross-sectional area is another.

** In fact, in the context of APM-level nanomechanical engineering, the term "molecule" will often refer to stable covalent structures that consist of fused rings. Lacking conformational degrees of freedom, these structures can hold definite and very robust shapes. There are molecules of many other kinds, of course, and some can be modeled as jointed, swiveling chains—the molecular strands in typical polymers, for example. Being far more accessible to chemical synthesis, folded polymeric chains are the molecules of practical interest today, in the context of current nanomechanical engineering (Chapter 12 and Appendix II explore the potential of these more accessible machines).

diamond, for example, consists of a single molecule, and so does the bulk of a rubber tire.

There's no point in trying to visualize something as being "invisibly small." Atoms, molecules, and red blood cells are all invisible, yet they differ in mass and complexity by factors of a trillion or more. Consistency in magnification is key to forming a useful, coherent visual image.

Imagine magnifying the nanoscale world by a factor of ten million, stretching nanometer-scale objects to centimeters. In such a world, a typical atom becomes a sphere about 3 mm in diameter, the size of a small bead or a capital O in a medium-large font. Simple atomically precise machine parts like nanoscale gears and bearings become a few centimeters long. More complex AP gadgets like nanoscale motors, planetary gears, and cam assemblies are larger, but still small enough to fit in the palm of your hand. If we look at the world through this magnifying lens, nanoscale and conventional machines become directly comparable in size.

Nanomachines and macromachines have similar sizes in this magnified view, but consider what happens to *micro*scale objects. In our ten million times magnified world, a human white blood cell just ten microns in size would fill a football stadium; a human hair of one hundred microns would become a cylinder a kilometer in diameter; and a magnified pair of human arms would become long enough to embrace the real-world Earth.

Conveniently, at this magnification a nanomechanical system as complex as a factory becomes roughly the size of a factory. Through such a lens, APM systems appear surprisingly mundane, a combination of things that are largely familiar with some aspects that are novel—different, but far from mysterious.

STRETCHING TIME
IN EQUAL PROPORTION

Now let's take a further step and stretch time in the same proportion, like using a slow-motion video to capture motions that would otherwise be too quick for the eye. The same factor of ten million that stretches nanometers to centimeters then stretches nanoseconds to centiseconds, and microseconds to about a minute and a half. This step—scaling time—is the key to a surprising result.

Stretching space and time in equal proportion scales properties like mass, force, and velocity in exactly the right way to make mechanical motions the same. One nanometer per nanosecond is exactly the same speed as one meter per second, just measured in different units, and regardless of size, if their speeds are the same, machine parts with geometrically similar shapes will trace geometrically similar paths in spacetime.

Consider two ordinary wheels of different sizes, both rolling along a road together. If one wheel has half the diameter and circumference of the other, it must turn twice as many times in order to roll the same distance—which is to say, it must rotate at twice the frequency. Thus, keeping rolling speeds equal, cycles per second scale in inverse proportion to size, and stretching space and time to make both sizes the same will make the rolling motions identical. This scaling law holds for any pattern of motion; it's simply a matter of geometric similarity.

Now consider two bells made of identical materials, one half the size of the other. When both are struck, the smaller will vibrate at twice the frequency. If you stretched space and slowed time by a factor of two, however, the smaller bell would vibrate in just the same way as the larger, with identical modes, tones, and frequencies. In other words, holding density and elastic modulus constant, vibrational frequencies scale inversely with size.

Finally, consider two strings made of the same materials, one half the length and diameter of the other. If both are stretched equally taut—halfway to their breaking points—they will both stretch in equal proportion. When plucked, the half-size string will emit a tone an octave higher than the longer string, which is to say, it will vibrate at twice the frequency, and once again, the frequency of motion scales inversely with size.

Our wheels, bells, and strings illustrate a beautifully simple principle. If objects are made of similar materials—equal in density, stiffness, and strength—their motions will scale together with space and time if speed and stress are held constant. The modes and frequencies of the bells, the dynamics of mass and force in the strings—even the stresses and strains caused by centripetal forces in spinning wheels—all these become identical when space and time are stretched to make sizes match in a magnified view. And what is true of wheels, bells, and strings is equally true of machines.

Thus, the surprisingly simple result is that, holding speed, stress, and materials properties constant, large and small machines behave alike. Reducing the scale of machinery by a factor of one-ten-millionth increases all motion frequencies—hence the number of operations completed per second—by a factor of ten million. In other words, the behaviors shown in our imaginary slow-motion videos are both qualitatively and quantitatively accurate.

This simple scaling relationship is key to understanding the world of nanomachines, an essential aspect of the basis for radical abundance.

Yet it's not quite enough for us to picture the size and speed of machines in the nanoscale world. To gain a firm grasp of the workings of APM, we should also be able to imagine how the components would *feel*. Before turning to future nanomachines, however, let's take a moment to look at the atomic texture of things that are more familiar.

VIEWING ATOMIC TEXTURES IN A MAGNIFIED WORLD

As the ancients suspected and chemists confirmed, all the different materials we find in the world consist of the same basic components. Though we may know on an intellectual level that atoms and molecules form everything we can touch—and even our bodies—the link between what we see and molecular properties isn't always obvious, yet on closer inspection, the links sometimes become clear.

Through our ten million times magnified view we can gain a perspective on the underlying nature of everyday things before turning our attention to the properties that enable nanomachines to behave in such radically ordinary ways. To get a sense of how familiar materials can reflect the characteristics of their atomic components, consider four very different things: water, a fishing line, a fish, and the gasoline engine of a motorboat.

Why is water fluid? The individual molecules aren't fluid, and at low temperatures they link together to form ice. Instead, water is fluid because its molecular parts (just a fraction of a nanometer in size) are able to move past one another, clinging together while changing partners, but not locking in place. Each molecule dances unpredictably, a consequence of thermal motion, yet on a larger scale water behaves so predictably that physicists often ignore individual molecules. Instead, they think and calculate in terms of *fluid* dynamics, governed by quantities like pressure, velocity, compressibility, density, and viscosity. Water can be described as a fluid down to a scale measured in nanometers before its atomic-scale texture forces a shift to thinking in terms of *molecular* dynamics.

At that nanometer scale, water looks very different. In our ten-million-times magnified view of the nanoscale world, we'd still see a kind of fluid mobility, but no longer smooth. If one could watch the molecular motions involved, the liquid surface would resolve into a pool of dancing, churning, vibrating beads (thermal motion again) that cling together yet sometimes break free to join the swarm of flying molecules we call "water vapor."

Next, imagine a fishing line made of nylon. Its most notable characteristic is its strength—the material holds together, whether slack or taut, along the whole length of the line. Its qualities are described with the vocabulary of solid mechanics, with quantities that parallel those of fluid mechanics, but with nylon's solid, elastic properties replacing fluid viscosity. As with water, the molecular texture of nylon becomes noticeable only at the scale of nanometers. Through our lens, it becomes clear that the strength of the macroscopic strand reflects the strength of chains of bonds that form the backbones of long polymer molecules, which may extend for hundreds of meters in our standard magnified view. Thus, a strong macroscale strand resolves into a bundle of strong molecular strands that cling together and hold one another in place. While each polymer molecule resembles a beaded necklace, slim enough that your fist could grip dozens of strands, in the magnified view the thickness of the fishing line itself would be as broad as a lake.

Though liquids and polymers can be pictured fairly easily, a far more complex picture emerges when we examine an organism—or even a single cell—in the nanoscale view.

There is nothing simple about a fish. Over the course of their day-to-day-lives, fish sense things, move, metabolize, repair tissues, and grow. As was true for simple fluids and solids, however, these macroscale activities reflect similar behavior on a smaller scale, activities within individual cells of particular kinds that themselves can sense things, move, metabolize, replace their parts, and grow. And looking more closely, one finds that these microscale, cellular behaviors emerge from corresponding activities on the molecular scale. Cells sense their environments through molecular devices (in the retina, for example, photon-sensing opsin molecules and hormone-sensing receptors in every tissue); cells move by means of molecular motors (such as myosin in muscle and dynein in flagella); cells employ enzymatic machinery (and ATP-synthase, a motor-driven piece of production machinery) to transform simple molecules in the course of metabolism; and cells replace their parts and grow by building larger, more complex molecular components using programmable molecular machine systems (and, in particular, ribosomes).

These molecular devices work in odd and unfamiliar ways, quite unlike anything we see in our macroscale world. Although some serve as machines, the motions of these devices are intimately tied to the thermal motion of the water molecules that surround them and they share more than a hint of water's fluidity.

Magnified ten million times by our lens, the molecular machinery within a cell is revealed to be a multitude of molecular components, each flexible and irregular in form, and yet structured with atomic precision. Although these molecules are often small enough to rest in the palm of your hand, some are linked together, forming assemblies that can range from tens of centimeters in diameter (the size of a ribosome) to tens or hundreds of meters in length (like the microtubules and fibers that make up the cytoskeleton, a mobile structure that gives cells their shape).

By peering this deeply into our fish, we have entered the world of molecular biology, the source of the molecular fabrication tools that enable much of today's progress in AP engineering. With their soft, fluid structures, cells are nothing like APM factories, yet their tools have proved useful in building molecular components of the kind needed to build new tools that can in turn build tools on the way to building APM-level systems, step by incremental step.

Advanced nanoscale tools will have much in common with today's macroscale machines, so it's worth looking more closely at those as well.

Alone among our examples, the nanoscale structure of a gasoline engine lacks distinct molecular components and shows little hint of its macroscale uses. The mobility of water stems from the mobility of its constituent water molecules; the strength of nylon strands stems from the strength of its constituent polymer strands; molecular motors enable organisms and their cells to move; molecular sensors enable eyes and their cells to sense light, and so on through many examples. But what of an engine and its metallic components?

Focusing on materials alone, we find that the properties of the steel in an engine's machinery do indeed reflect properties of metals at the molecular scale, the properties of so-called "metallic bonding" between layers of atoms. If you could watch what happens inside a sheet of steel when it's struck by a hammer, you'd see how layers of atoms in metal can slide over one another, allowing steel to yield under stress; a sheet of glass, by contrast, has no layers that can slide and yield, and so where a steel sheet would dent, a window will shatter.

The dynamic power of an engine as a whole, however, has no parallel in the molecular-scale properties of steel, nor in its lubricating fluids, nor in the blind release of heat when fuel molecules combine with oxygen. While a muscle transforms molecular energy into motion using molecular machines, an engine does nothing of the sort. Its macroscopic function is a purely macroscopic effect and the nanoscale structure of an engine shows no engine-like behavior.

Developing the mechanisms required for advanced APM will involve translating the functional patterns of macroscale machinery into the nanoscale world. The functions of rigid metallic structures, like the gears and shafts of an engine, can be performed by rigid, atomically precise nanoscale structures, albeit with several key differences. For example, because heat flow doesn't scale the same way as mechanical motion (in a nanoscale container, gases cool almost instantly), combustion must give way to non-thermal methods of harnessing chemical energy, and likewise, because electromagnets don't scale the same way as mechanical motion (with thinner wires and smaller currents, magnetic forces become too weak), electric fields naturally replace magnetic fields in electrical motors. As it happens, however, these alternative ways of harnessing chemical energy and electrical power can outperform current engines and motors, substantially upping efficiency while increasing the potential power densities to levels constrained primarily by macroscale cooling capacity.*

Replacing internal combustion engines with APM-based energy conversion technologies would complete the symmetry noted above. The function of a macroscale engine would at last be reflected at the molecular, nanoscale level, bringing engines in line with fluids, fishing lines, and fish.

THE TEXTURE OF ATOMICALLY PRECISE MACHINE SURFACES

Neither nylon, nor steel, nor fish contain parts that resemble machines as we know them today; even biological molecular machines have an alien character, being made of flexible, irregular, polymeric components moved by thermal motion. Nanomachines for advanced APM will have an unusual texture, yet will be familiar in terms of form and function.

Indeed, the functions of these nanomachines can seem almost absurdly familiar, but for good reason. In a factory, a machine's tasks of moving and positioning parts for assembly are essentially independent of scale, and mechanical scaling laws suggest that these tasks can be accomplished in much the same way by systems of nanomachinery.

---

* Mechanical scaling laws (high frequencies again) enable these nanoscale devices to achieve high energy throughput by means of motions that carry molecules and electrons across nanoscale distances in times measured in nanoseconds.

What is different in the nanoscale world, however, is the way in which materials interact when they touch—an important point, as a machine can't work well if its parts can't move smoothly, and interactions between atomic-scale bumps on surfaces might seem to make smooth movement impossible.

In machines, most moving parts have surfaces that either press together without slipping or slide across one another with minimal friction. Bearings, for example, support motors and drive shafts on rolling balls or cylinders, or simply use smooth, lubricated surfaces that glide over one another; in bearings the aim is to slide with minimal friction. Gears and drive belts, however, require the opposite, surfaces that touch or mesh without sliding at all.

On the nanoscale, there can be no truly flat surfaces. If magnified ten million times, even the flattest, smoothest surface reveals itself to be an array of closely spaced bumps, each bump an atom. To understand how nanomachines can function with bumpy surfaces, consider what happens when atomically textured surfaces touch.*

Suppose you were to touch a flat surface, textured with densely packed rows of atoms, while wearing a glove with atom-textured finger tips. Just before touching the surface (within a millimeter or so), you'd feel your finger pulled down to snap into contact. It's the same sort of sensation you get when you try to bring a strong magnet in contact with a refrigerator door. When the magnet comes close enough to the surface, it suddenly snaps down sharply. The effect here is similar, though not caused by magnetism; the cause is an inherent, short-range attraction that acts between all forms of matter, variously called van der Waals, dispersion, or London dispersion forces (the famous Casimir forces are a weaker, longer range aspect of the same interaction).

With your fingertip pulled down into contact, you'd feel a surface almost as hard as the material itself (about as soft as a slab of steel). If your fingers were very sensitive, however, you could feel just a little give where the bumps are in contact, as if their surfaces had a thin, soft layer a frac-

---

* The following remarks apply only to strong materials with unreactive, nonpolar surfaces. Surface scientists may want to picture graphene, or a passivated, covalent surface like hydrogen-terminated diamond or silicon.

tion of a millimeter thick. This slightly soft resistance to pushing one atom into another, called "overlap repulsion," is the same force that keeps you from walking through walls.

To get a sense of how this force feels, zoom in for a moment to make atoms the size of Ping-Pong balls, and grasp one in each hand. Each atom would feel like a hard sphere wrapped in a soft, slippery layer a few millimeters thick, and as you pushed two spheres together and moved one past the other, they would feel like magnets repelling each another, keeping their distance without touching or rubbing together, each forming a bump in a frictionless force field.

So now (with all atoms returned to their standard size), your fingertip is resting on the surface, and both surface and glove are textured with columns and rows of small, smooth, atomic bumps. If the glove were fully flexible, you could feel individual bumps, but instead, imagine that the glove's fingertip is as hard as steel and only slightly curved (almost flat). Now, sliding your gloved fingertip across the surface, you wouldn't feel the force of individual bumps pressing against your finger, but instead the sensation of one array of closely packed bumps sliding across another. This is where the results start to interest a machine designer.

———

WHAT IF YOUR MACHINE needed a pair of surfaces that slide or a pair that could roll without slipping? If the surfaces had irregular structures, their atomic-scale bumps would make sliding or rolling difficult, but surfaces with regular, atomically precise rows can serve the purpose, depending on spacing and alignment.

First, consider what can happen when the rows are aligned, so that the ridges formed by the rows of atoms on your glove rest in shallow grooves between the rows on the surface. Sliding your finger across these rows, the surface would feel strongly corrugated. If the ridges and grooves were meshed together more deeply, you'd find it impossible to slide your finger across them. There'd be no tendency to slip, but the curved surface on your fingertip would allow you to rock your finger back and forth: the meshing ridges would act like the teeth on a gear, and in a machine they could serve exactly that function. Gears are easy when surfaces mesh, but what about making sliding surfaces for bearings? Turn your fingertip at an angle, so that the rows can't mesh, and once again try to slide it over the surface. Now, each row slides across many others simultaneously, and because individual atoms are like soft, slippery spheres, when the rows don't mesh, you'll feel little resistance to sliding. In fact, if the mismatch is just right, you'd feel scarcely any bumps at all, and computational models then predict extremely low friction.

Experimentalists have confirmed this prediction. In laboratories the tip of a scanning probe microscope plays the role of a finger and a flake of graphite or molybdenum disulfide plays the role of a glove, and sliding the tip across a surface tests the frictional forces. Scientists have dubbed this phenomenon "superlubricity" and called it, with only slight exaggeration, "a state of vanishing friction." By taking advantage of superlubricity, nanomachines can operate with nearly frictionless efficiency.

This mismatching-rows trick can be made to work in other ways, including not only mismatched angles, but mismatched row spacings and mismatched numbers of rows around the circumferences of nested cylinders. Using electron microscopes to observe how nested carbon nanotubes move, experimentalists have demonstrated each of these principles in making and testing both linear and rotary bearings. Again, the results match computational models and confirm key principles of operation for advanced nanomachines.

Nonetheless, some frictional forces remain. Although static friction can be effectively zero, there's still a resistance that grows in proportion to speed. This is a drag force, like the resistance you'd feel in the macroscale world when sliding a finger quickly across a surface lubricated with a very slippery but slightly viscous fluid.

Rolling without slip; sliding with almost no friction—these are the properties that engineers typically want for moving parts in contact. (A third is sliding *with* friction, to serve as a damper or brake, but as one might expect, adding friction is easy.)

With rigid solids and surfaces that can slide or roll (together with a few other straightforward components, like springs and brakes), almost the whole of mechanical engineering follows with surprisingly few differences other than scale. Thus, most of what we understand about machines and machine systems—familiar parts of the modern world—carries over to the design of nanomechanical systems that can do the work needed for atomically precise manufacturing. And because the scaled view is both qualitatively and quantitatively accurate, informal understanding and engineering analysis coincide.

Where are the differences, though? So far, the behavior of machines at the nanoscale has been quite familiar—we've had only to make allowances for atomic texture—but there's another aspect to consider. Before we can leave our ten-million-times magnified view of the nanoscale world, we must look more closely at a phenomenon that doesn't affect macroscale machines: thermal motion.

At any temperature above absolute zero, thermal energy drives seemingly random molecular motions. This thermal motion is an unavoidable nanoscale phenomenon and it doesn't scale the same way as mechanical motions. Although negligible at a macroscopic scale, thermal motion becomes an important consideration in nanomechanical engineering.

What does thermal motion look like in our standard, scaled view of nanomachines? The short answer is that it's too fast to see. Molecules in the air move at about the speed of sound. In our magnified view, with space and time both stretched by a factor of ten million, speeds are unchanged, and air molecules would be impossible to see, like bullets in flight. Although atoms move at similar speeds in solids, they appear to be stationary, because (at more than a megahertz, even when with scaled time) their small amplitude vibrational motions are invisible.

Thus, on the rigid surface of a nanomachine component, moving atoms would show no more than a slight motion blur, with amplitudes a few times the diameter of a human hair. It's often useful to think of these motions as slightly changing the surface texture. Thanks to this ceaseless vibration, a soft, slippery molecular surface would feel just slightly softer and slightly more slippery.

Each machine component will quiver in place as a whole and the amplitude of its motions will depend on how firmly it's linked to the rest—which is to say, the stiffness of the elastic restoring forces that resist its displacement. Large-amplitude thermal motions can cause manufacturing machines to make errors in where they place parts, but adequate stiffness (and geometrical tolerances) can make such errors extremely rare. In typical, reasonably rigid machines, the mean thermal amplitude can be restrained to less than a tenth of an atomic diameter, enabling highly reliable operations.

THERMAL MOTION RULES CHEMISTRY AND MOLECULAR BIOLOGY

In machines thermal motion is sometimes important, but in liquids and biomolecular nanomachines thermal energy is always *overwhelmingly* important. Thermal motion is responsible for the tumbling, wandering diffusion of molecules that makes liquids behave like liquids, and solution-phase chemistry relies on thermal motion not only to bring molecules together, but also to kick molecules over energy barriers that impede chemical reactions.

What does solution-phase chemistry look like in our standard, magnified view? In short, incomprehensible. First, consider the solvent, which might be water. When I described water earlier, I cheated a bit in saying that the fluid would look like a pool of dancing beads. In fact, moving at sonic speeds, the molecules would disappear into a motion blur; to make them visible would require slowing them down by a factor not just of ten million, but of more than ten billion.

Now consider the chemical reactions themselves, the changes in bonding that can occur when two molecules meet in solution and happen to have enough thermal energy to rearrange bonds. In the macroscale world of a laboratory, reaction times can vary enormously, but a reaction time of around a minute is often considered conveniently fast. But what would you see if you could follow a single reactive molecule, slowed down enough to enable the eye to follow its motions? You might not want to try, because to have good odds of seeing a reaction occur, you'd have to watch for about a billion minutes, which is to say, about two thousand years.

Thus, the scaled view that gives such powerful insights into the behavior of rigid nanomachines is essentially useless for understanding solution-phase chemistry. Chemists must work in a different conceptual framework, one seldom concerned with mechanical speed, force, or position.*

Thermal energy also dominates motions in biomolecular nanomachines, devices which straddle the gap between the chaos of solution-phase chemistry and the orderly motion of gears, bearings, shafts, and the rest. Many biomolecular components are made of protein, which makes proteins worth a closer look.

To imagine a protein at work in water, first make the water molecules transparent to clear away the surrounding solution-phase blur. Magnified ten million times, a typical protein would be a few centimeters wide and might be part of a ribosome, a structure several times larger and many times bulkier. Computer graphics of proteins show that they're peculiar objects, looking much like a bunch of grapes, each representing a highly magnified atom.

Their surfaces have floppy protrusions, parts of the side-chains of amino acids, the phenyl of phenylalanine, for example. At our standard magnification of space and time, the floppiest side chains would blur like water molecules, while others would be packed together to make a stable, solid interior. If the protein itself is free to move, it will tumble and shift its location from moment to moment.

Machines built out of proteins in the style evolved by biology have further unusual features. Powered by chemical reactions (or flowing ions, or light), proteins interact with smaller molecular objects at the speed of chemistry—in other words, with random motions too fast to see, punctuated by significant events that may only rarely occur.

Consider the protein machines that move your muscles. These consist of myosin molecules that take steps along actin fibers, each step moving forward by thermal motion. If we were to peer through our nanoscale lens at a leg muscle in the midst of a macroscale step, we would see myosin molecules thrash in a blur, moving forward in nanoscale steps seemingly by happenstance, with energy from ATP molecules applied, not to drive steps forward, but to prevent steps backward.

---

* Fundamental chemical principles are universal, of course, and apply perfectly well to mechanically guided molecular reactions, provided that one takes care to describe mechanical constraints in chemical terms. Some solid-phase, enzymatic, and intramolecular reactions provide good models.

Likewise, if you watched a ribosome at work building a protein, you'd again see a thrashing blur as the ribosome worked, cleaving and forming bonds, binding and releasing tRNA molecules, and adding amino acids to the chain at a rate of about one unit per week, yielding a new protein after several years of scaled time (in real time, just a few seconds).

Soft machines like these, made of folded polymers, driven by thermal motion, working in water, fed parts by diffusion, are different in kind from nanoscale machines made of rigid materials, driven by motors, working without fluids, and fed parts by other machines. Biomolecular materials, however, can be used to build machines that work in more comprehensible ways, machines that make non-biomolecular products that themselves can serve as parts of next-generation machines. The path forward from there leads to APM-level technologies, not all at once, but through incremental steps in materials, design, and production capabilities.

WHY APM SYSTEMS ARE SURPRISINGLY ORDINARY

Looking back on this view of the nanoscale world, we can see that some aspects—the ones that matter most to nanomechanical engineering—are surprisingly ordinary. Selected materials in selected configurations can behave as machines. Atomic texture may be unfamiliar, but isn't hard to imagine, while thermal motion can be tightly constrained, quite unlike the large, random motions found in molecular biology and solution-phase chemistry.

Nanomachines made of stiff materials can be engineered to employ familiar kinds of moving parts, using bearings that slide, gears that mesh, and springs that stretch and compress (along with latching mechanisms, planetary gears, constant-speed couplings, four-bar linkages, chain drives, conveyor belts . . .). Machines made of parts like these can perform the functions of conventional machines, but with their motions scaled smaller in both space and time.

This is why the machinery of APM systems can be so radically ordinary: The motion tasks in manufacturing are independent of scale and can be solved by similar kinds of machinery. Parts must be held and moved, transferred from machine to machine, rotated, positioned, and fastened together.*

In the end, what stands out is that advanced nanomachines can closely resemble the machines that have enabled the Industrial Revolution. From an engineering perspective, an enormous mass of knowledge can be transferred, adapted, and applied from the machines of the past to the machines of the future. From anyone's perspective, our shared intuitions about objects like tools—things that we can grasp with our hands and our minds—can help us grasp the nature of the machines that will enable the APM revolution. The scaled view that shows nanomachines as familiar, understandable things—a view magnified ten million times in distance and slowed ten million times in time—provides an intuitive picture that is both qualitatively and quantitatively accurate.

---

*Note, however, the self-assembly in fluids—driven by thermal, not mechanical motion—is the method of choice for molecular assembly today, and that as Appendix II discusses, thermal and machine-driven motion will work well together across a spectrum of intermediate technologies: There is no gap between self-assembly and loosely constrained, machine-guided methods.

CHAPTER 6

The Ways We Make Things

NOW THAT WE'VE EXAMINED the texture and behavior of objects at the nanoscale, we need to consider the nature of APM itself—how it works as a production technology. I have described it as a fourth technological revolution like, and yet unlike, the Agricultural, Industrial, and Information Revolutions that have reshaped the human world. But what is APM really like as a production technology, compared to the technologies we use today?

Our journey so far has viewed APM from several very different perspectives. A reader who focused only on the general engineering principles might view APM as if it were a kind of computing, with atoms in place of bits, or perhaps like ordinary manufacturing, but based on smaller machines. Another reader might focus on the molecular aspect, and see it as closely akin to chemistry or biotechnology.

These perspectives are useful, but a closer comparison to today's production technologies can put APM in sharper focus. Looking back on the picture so far, the broad, useful comparisons are these:

- Like conventional manufacturing systems, APM systems assemble small parts to make larger parts and products.

Though operating at vastly different scales, both kinds of manufacturing use comparable machines to move parts from one place to another and put them together.

- Like digital information systems, APM systems depend on nanoscale components to construct complex patterns from the simplest possible parts and can employ comparably precise, reliable steps to produce precise, reliable results.
- Like biomolecular systems, APM systems build complex structures by assembling molecular building blocks to form patterns defined by digital data. Though APM and biomolecular systems differ profoundly in terms of structure and operational principles, they both use nanoscale machinery to guide molecular motions.
- Like chemical synthesis, APM systems assemble larger bonded structures from smaller reactive molecules. Machine-guided motion offers more control than thermally driven diffusion, yet both processes rely on identical principles of molecular physics.

Each of these areas illustrates aspects of APM, but a closer look from a fresh angle can shed light not just on APM, but on the material foundations of modern civilization.

PARTS, MACHINES, AND THE POWER OF AUTOMATED MANUFACTURING

Conventional manufacturing begins with machines that shape materials to make parts; APM begins with machines that move molecules, where the molecules themselves are the parts. Parts shaped from materials in conventional ways—whether molded, hammered, drilled, or cut with a lathe—can never be shaped with complete precision. Molecules, however, start out atomically precise and if each succeeding step produces a specific result, the products will retain this initial precision.

To put parts together, the motion-control tasks of machine assembly are in essence the same, regardless of scale. A machine must be able to transport streams of parts from place to place and put them together to form larger assemblies on the way to becoming completed useful products. To build machines that perform these motions, engineers have accumulated a kit full of design solutions, including devices like conveyor belts, rotary feeders, pick-and-place mechanisms, continuous-motion assembly machines, and (more famous and photogenic) the jointed, arm-like devices called "industrial robots."

Many factories today use both machines and human labor to perform assembly, and economics largely determines the extent of machine automation. Machine-based assembly has been in practice for decades (in the automobile industry, for example) and industrial robots are gaining ground as their costs fall and capabilities increase.

APM can deliver macroscopic products (built from smaller parts, in turn built from yet smaller parts, and so on), and the macroscopic end of the process can be almost perfectly conventional. Indeed, an APM-level technology could change the world simply by delivering advanced-technology, precisely manufactured parts for assembly in a fully conventional factory, and it would be natural for the revolution to begin that way.

Looking back from the end of a product assembly line in a conventional factory, one sees a stream of finished products emerge—but where do the parts come from? They're all products of streams of converging processes. Each tributary stream, if traced back to its origins, begins with machines that make the smallest-scale parts, shaped from materials like plastics, ceramics, glass, and steel.

In an APM factory, this familiar architecture of converging assembly paths could be traced back through many more stages, leading to machines that handle nanoscale molecular parts. These machines can be designed in a similar way back to the smallest sizes. In the previous chapter, we've already seen that even nanoscale machines can closely resemble those built on a larger scale, aside from performing nanoscale motions quickly, in nanoscale time.

In a mechanical, architectural, and abstract sense, today's factories provide a template for APM systems. Indeed, a key aspect of radical abundance—radical productivity—follows from what this template implies in conjunction with simple scaling laws.

Machines scaled down by a factor of ten million can perform ten million operations in the time it would take a similar macroscale machine to perform just one operation. Thus, in manufacturing, scaling machines by a factor of ten million translates directly to a dramatic increase in physical productivity as measured by the mass that can be processed per second by a given mass of machinery; at the nanoscale end of the process, this scaling principle is the basis for high-throughput APM.

Manufacturing Without Moving the Parts: Nanolithography for Chips

One branch of manufacturing has already entered the nanoscale world, applying specialized techniques to fabricate devices with billions of parts already in place (no assembly required). This is, of course, chip manufacturing, which uses lithographic techniques that, in essence, resemble conventional printing but work with different materials at a much smaller scale. Through visible-light photolithography, printers can lay down patterns of ten million pixels; using ultraviolet-light photolithography, semiconductor manufacturers lay down patterns that equate to billions of pixels, and on a chip far smaller than a printing plate.

Then, in processes similar to developing an image on photographic film, manufacturers use chemical methods to bring out these patterns, not as light and dark images, but instead as contours in a plastic film on a silicon wafer. By exposing these patterned surfaces to reactive gases, heat, metal vapors, and ion beams, semiconductor fabrication machines etch, deposit, and oxidize materials that form the layers of chips—transistors, insulators, and strips of metal that serve as wires.

Lithographic patterns aren't atomically precise, and they can't be, because light and chemical exposure can't provide the necessary degree of control. And today, as exponential Moore's Law miniaturization approaches the atomic scale, the challenge of increasing precision has begun to flatten the curve of progress.

When compared to the mechanically driven, atomically precise operations that define APM, semiconductor lithography shows more contrasts than similarities. It is telling that semiconductor fabrication
facilities are huge, housing arrays of machines that can cost billions of
dollars, while molecular fabrication today—with full atomic precision,
beyond the reach of even high-resolution photolithography—is often
done by university students using tools like pipettes and glass beakers.

A Special Manufacturing Method: 3D Printing

Another emerging method for manufacturing also breaks the pattern of
making and then assembling parts: 3D printing, sometimes known as
additive manufacturing.

3D printing differs from the traditional ways of shaping materials.
Some traditional methods make a shape all at once using a costly, specialized tool, like a mold to shape plastic, a die to stamp steel, or an optical mask in semiconductor lithography. Other traditional methods
carve shapes by removing small bits of material using general-purpose
equipment like lathes, drills, and milling machines. 3D printing, by
contrast, makes shapes by *adding* small bits of material using general-
purpose machines guided by digital data files. 3D printing can make
shapes beyond the reach of casting or carving.

Today, the most common 3D printers produce structures by moving
a nozzle as it lays down patterns of melted plastic in much the same way
that ink-jet printers make 2D images by moving a print head as it lays
down patterns of ink.

3D printing technologies can do more than this, however, and are
advancing rapidly, with falling costs and an expanding range of materials
and products. Although costs remain high by mass-production standards, 3D printers offer matchless flexibility for making unique or intricate objects. At the high-cost end of the market, some machines make
metal objects with unprecedented freedom of form, for example, customized titanium jaw implants for reconstructive surgery. Meanwhile,
in industry, aerospace companies see potential applications for engine
components with optimized forms and organically sculpted titanium

manifolds. Printable materials now include not only plastics and metals, but ceramics, vascular tissues, and chocolate.

Meanwhile, at the low-cost end of the market, 3D printers have found their way into homes, where their users make toys and small sculptures, multi-part gadgets fabricated in a single pass with gears already in place, and custom-made components, like brackets, for practical use. As 3D printer technology spreads, so do online user communities. Users share open-source digital design files as freely as others share digital images, and the RepRap community, for one, uses home-built machines to make customized components for new and better machines.

Like APM systems, 3D printers build physical objects from small bits of material and the machines themselves are compact and affordable. Despite their radical differences in cost, throughput, and product range, 3D printers give at least a hint of what APM systems will enable and a hint of how human communities may share information that can be translated into physical, functional forms.

BITS, ATOMS, AND THE POWER OF DIGITAL SYSTEMS

It may seem odd to compare the processing of information with the processing of matter, yet the parallels between today's digital computing and tomorrow's APM-based production run deep. At the finest-grained level, the discrete, atomic structure of matter enables APM systems to work with patterns of atoms much as digital information systems work with patterns of bits. What's more, the shift from analog to digital systems in electronics has strong parallels with the prospective shift from conventional manufacturing to atomically precise production.

Seen through our ten million–fold magnifying lens, a single nanoscale transistor in a present-day chip would be as large as a tablet computer. Viewed through the same lens, a computer chip would consist of a billion transistors buried deep under wiring that would stretch to the horizon. Indeed, an entire computer chip would be about the size of Belgium.

As we've seen, today's chip fabrication methods fall short of atomic precision. Our magnified transistor, the size of a tablet computer, would be about as rough as a block chopped from wood with an axe. Today's nanoelectronics is crude, in a physical sense, the result of building *digital* information systems with what is, in effect, *analog* manufacturing.

The step from analog manufacturing to APM is like the step from analog to digital electronics. Despite the difference between information and matter—between bits and atoms—the digital information revolution and APM draw on similar principles and sources of power.

Analog electronic systems represent and convey signals by physical gradations in some medium, gradations that are analogous to gradations in something else—the amplitude of a radio signal, for example, representing the form of sound waves in a human voice. Vinyl records work in a similar way, employing rippling grooves stamped into plastic to convey past patterns of sound. Analog components and circuits, though never quite perfect, were engineered with an artistry honed by decades of design and experiment.

Digital electronics has little in common with its analog relatives. Where analog systems rely on physical gradations, a digital system relies on a series of numbers encoded as bits. A description of sound waves, whatever their form, will be encoded using only two symbols: 0 and 1. This symbolic description, like text, reads the same no matter how it's written—whether in large print, small print, or block letters in crayon. The distortion inherent in analog playback is no longer a concern. Exact, unambiguous results, in effect, define what it means to be "digital," and this principle ensures the crystalline clarity of digital radio.

Because of the distinct, right-or-wrong nature of symbols and the simplicity of a system founded on only two digits, even crudely made devices can process digital signals with absolute perfection. This is what enables nanoelectronic systems to operate reliably using devices that themselves are far from precise. If nuts and bolts were misshaped by 5 percent, they wouldn't fit together, yet a present-day chip, when built with twenty nanometer devices subject to nanometer-scale imperfections, tolerates errors of similar proportion.

The key to this performance is the digital principle (as noted before), which allows a range of voltages to be treated as "1," provided that this range is well-separated from another range treated as "0." By designing circuits that keep signals well within their proper ranges, limited errors in signals—including thermal electronic noise—have no effect on how the output is read. There's a margin for error. With careful design, molecular operations guided by nanoscale machines can be made to work in a similar way.

Atoms aren't as simple as bits, yet like bits they can form perfectly distinct patterns. In a digital information system, a bit is either a 0 or 1, while in an APM system, an atom is either bonded in precisely the right place or it isn't. In digital electronics, this all-or-nothing discreteness must be engineered into the circuits, but in APM, discreteness is inherent in the atoms themselves and in the very nature of typical covalent bonds. Deep down, the discrete nature of quantum mechanics enables precise reproducibility, and different patterns of atoms can be as distinct as different patterns of bits.

In digital electronics a noise margin is the voltage gap that separates a reliable 0 or 1 from an ambiguous value, but in APM a noise margin results from a physical gap and a mechanical restraint. Roughly speaking, directing a molecule to the site where a reaction should occur is like placing a peg in a chosen hole while being jostled by random vibrations. If the molecule were to encounter a different yet similar site, it could bond where it shouldn't, like a peg sliding (and locking) into the wrong hole. Whether directing a molecule or inserting a peg, the key to avoiding an error is to resist shaking forces that could cause a misplacement, missing the intended target and hitting another instead.

Thus, in an APM mechanism, thermal motion causes unavoidable, random fluctuations in position, and constraining thermal fluctuations tightly enough to avoid misplacement is the key to reliability. What this means in practice is that a machine must be stiff enough that bending it far enough to result in misplacement requires more energy than thermal fluctuations can likely supply. This resistance to bending is an elastic restraint, and the energy required to cause a misplacement (the energy barrier) provides a noise margin. To be more exact, to ensure high reliability the energy barrier must be dozens of times larger than the characteristic thermal energy, a quantity known as kT.

In digital electronics, a computer chip can be engineered to use billions of transistors to perform many trillions of device operations per second, all without error. Well-chosen APM operations, guided by well-designed mechanisms, can be made equally reliable.

Turning to prospects for future computers, digital information and digital fabrication will surely converge. Progress in digital electronics has led manufacturing deep into the nanoscale size range, and this line of advance will eventually demand atomic precision, yet this must be achieved by radically different means. When a device is no more than a few atomic diameters in size, each atom must be in the right place. Continued progress in digital systems will require this precision, and APM will enable it.

BIOTECHNOLOGY AND THE POWER OF MOLECULAR NANOMACHINES

Biological molecular machines today provide a working example of machinery for atomically precise fabrication of atomically precise nanoscale devices that make atomically precise nanoscale structures. This close parallel with APM systems makes biomolecular systems a rich source of analogies—some illuminating, but some deceptive.

Nature developed the first digital information systems and used them to direct atomically precise fabrication. Textbook biology teaches us that DNA encodes digital data in sequences of four kinds of monomers—nucleotides—each bearing two bits of data and read in blocks of six-bit words that encode the twenty standard amino acids found in proteins. Ribosomes read this information, produce the specified proteins, and thereby build the molecular components and devices that are so central to the workings of everything else in the cell, including the ribosomes themselves.

From a distance, a comparison between biotechnology and APM would appear to be quite fruitful, and it is, if the analogy isn't taken too far. At their bottom-most levels, APM systems have much in common with cellular metabolism in terms of function, efficiency, and molecular precision, yet they have little in common with cells in terms of their molecular structures or overall organization; biological cells are far too fluid and messy to have much resemblance. Experience shows that relying on biological metaphors can lead to far more confusion than enlightenment.

As we've seen, the orderly mechanical structures of APM architectures have far more in common with factory systems than they do with anything found in biology. Thus, while it makes sense to imagine nanoscale factory machinery, it's hopeless to try to imagine a mechanized cell—much like trying to imagine mechanized noodle soup.

The contrasts are pervasive. Factory machines, regardless of size, consist of rigid components that move in familiar mechanical ways; biomolecular machines, by contrast, are soft, irregular structures that move on jittery, unpredictable paths driven primarily by thermal motion. Likewise, factory machines receive and process parts in an orderly way; biomolecular systems typically bind and release randomly moving molecules. These differences are echoed at all levels of organization, from the contrast between rigid factory walls and fluid cell membranes to the contrast between machine-guided construction and processes like self-assembly and cell division.

Historically, biomolecular systems provided inspiration and a kind of existence proof, showing that intricate nanoscale mechanisms can be built with atomic precision by similar mechanisms and that construction can be guided by digital instructions. On the practical side, genetic engineers have learned to program biomolecular machines, providing a straightforward way to build complex, atomically precise structures according to design.

The most attractive paths toward APM, in my judgment, will extend current techniques for using or imitating biological methods—but at the level of components, not systems. Engineers using biomolecular materials need not merely imitate natural patterns. In structural DNA nanotechnology research, for example, nature's information molecules are being used to build large, intricate, structural frameworks. Some potential applications of these frameworks do involve information, but not in any biological sense. Researchers instead aim to use DNA structures as circuit boards for building digital electronics.

DNA, RNA, and ribosomes are biology's means of building protein machines, but enzymes (which aid these machines) perform a simpler function. Enzymes are the protein machines that catalyze specific chemical reactions in every step of digestion, metabolism, and molecular fabrication in cells. They speed and direct transformations by binding molecules and helping to guide their reactions, smoothing pathways that lead to particular products. In so doing, they can increase reaction rates by factors as high as a thousand (and sometimes a million).

Even outside biology enzymes can be useful devices. Protein engineers have tweaked enzymes to improve their capabilities, making enzymes more stable for industrial use, for example, to help produce fuels from cellulose waste, and even in homes, where enzymes engineered to survive in heat and detergents end up in washing machines.

In physical terms, a great distance separates enzymatic reactions from advanced APM, yet no sharp line stands between them. Instead, one finds a series of incremental steps that lead through ongoing advances in biopolymer engineering, advances that can provide a toolkit for guiding the fabrication of incrementally more robust materials and machines. Every step forward in this direction enriches the palette of materials for engineering design, and every step can improve the performance of tools that enable further advances.

We've already moved far along this path, and as prospects become clearer, the pace will quicken.

ATOMIC PRECISION THROUGH CHEMISTRY

Chemical synthesis provides an approach to atomically precise fabrication that is quite different from biomolecular machinery, one that provides a complementary range of products. The methods of organic chemistry aren't widely understood (even within the molecular sciences), and there's nothing like them elsewhere in science or technology.

Chemists have built structures with atomic precision for more than a century, and over the years their capabilities have grown in a way that resists any brief summary. The state of the art in semiconductor fabrication has its complexities, but one simple metric—the shrinking size of transistors—summarizes much of the story. Chemical synthesis, by contrast, has no comparable metric, no simple physical measure of overall progress.

To get a good sense of the nature of the field requires understanding both general physical principles (of bonding, thermodynamics, statistical mechanics) and the particulars of molecular functional groups and how they react. To get a good sense of the state of the art in the field—its current means, ends, capabilities, limitations—requires paying attention to the stream of results in major scientific journals, supplemented with the backstory of research life that comes only from laboratory visits and evening conversations with chemists.

In the deepest sense, what makes chemical synthesis different from other production methods is its radically different methods of controlling the fabrication process. Manufacturing processes put patterns together with direct control of where the parts go (even in printing and lithography), and digital electronics exercises similarly direct control of patterns of bits. Biomolecular machines blend direct control (guided motion in ribosomes, for example) with control based on how parts fit together and bind when random motion in fluids brings them into contact, like magnetized puzzle pieces snapping together; designing parts like these that bind in unique ways isn't direct control of assembly, but comes close.

Traditional chemical synthesis, by contrast, relies on far less selective kinds of molecular fitting and binding (i.e., chemical reactions) which work with smaller, simpler molecules that lack the distinctive puzzle-piece shapes found in biology. Chemists leverage this limited means of control by choosing the sequence in which reactive molecules are allowed to meet. In other words, they add ingredient A, then ingredient B, then ingredient C (and heat, distill, precipitate . . .) to make reactions occur in specific sequences. Along the way they often must block and unblock potentially reactive sites to control what happens where on a molecule. Each reaction is like a move in a game, and indeed, organic synthesis has been compared to chess. And as in chess, to become a grand master requires a lifetime of learning through study, failure, changes of tactics, and what is often hard-won success.

Although the specific molecular products of synthesis are, by definition, atomically precise, error rates along the way are usually high because by-products are usually common. Every step in a synthesis results in some percentage of loss and declining net yields limit the number of practical steps.

From a chemical perspective, the greatest value of mechanically guided molecular motion lies not in facilitating reactions by bringing molecules together, but in preventing unwanted reactions by keeping potentially reactive molecules apart. This reliable, general-purpose means of protection enables unprecedented reaction specificity, suppressing by-products and enabling long, high-yield reaction sequences. These sequences (which now can safely include highly reactive species) can be used to produce large, intricate structures of unprecedented kinds, including densely bonded, precisely structured covalent solids.

Suffice it to say that chemists have learned how to produce a range of molecules that, aside from practical limits on size and complexity, includes and surpasses the range found in biomolecular systems or, in fact, anywhere else in nature. These novel synthetic molecules can complement or supersede biomolecules as a basis for building complex and functional molecular systems. Thus, chemical synthesis will serve as a powerful tool as we advance further into the realm of atomic precision.

MECHANICAL, DIGITAL, MOLECULAR, AND PRECISE

As seen in this survey, human beings have developed many ways of making things in the years since the Industrial Revolution, and these ways have many connections with APM:

- Automated manufacturing shows us how to architect systems of programmable machines that can assemble large, complex systems from smaller components; as a consequence of scaling laws, similar machines can enable high-throughput APM systems.
- Digital information systems show us that nanoscale systems can perform intricate sequences of high-frequency operations and illustrate engineering principles that can enable nearly perfect reliability.

- Biomolecular systems show us that molecular machine systems are capable of constructing complex, functional AP components; indeed, through this capability, biomolecular systems have already enabled progress on paths that lead toward APM-level technologies.
- Advances in chemistry show us additional ways of producing AP molecular structures and illustrate how a wide range of structures can be made simply by bringing reactive molecules together; chemical synthesis, too, is enabling progress on paths that lead toward APM-level technologies.

Taken together, today's ways of making things offer insights into APM systems and how they can be implemented. What are the gaps that remain to be filled?

AP fabrication today lacks any broadly applicable means of guiding the motion of molecules with adequate precision to achieve tight control of outcomes, and instead relies primarily on chemical and biomolecular techniques. While chemical techniques are general but unreliable, bio-molecular techniques are reliable but highly specialized.

Looking forward, I see two crucial lines of development, the first well underway, but the second now barely begun:

- First, improvements in methods of designing large, functional molecular systems and the use of these methods to build simple machines from components of the sort made today by biomolecular and chemical means.
- Second, using simple machines of this sort as tools for building components made of materials that offer better performance and enable more straightforward design of next-generation machines.

What will faster progress require? Perhaps surprisingly, current knowledge and capabilities aren't the key constraints. What's missing isn't basic science or better means of making things (though both have great value); the challenge instead is to develop a better understanding of how to apply system-level engineering methods in the realm of molecular science and technology, to fill a gap in concepts and organization.

The next section will turn to this topic, exploring the sometimes uneasy relationship between the disciplines of engineering and science. They're more different than most people think.

PART 3

EXPLORING DEEP TECHNOLOGY

CHAPTER 7

Science and the Timeless Landscape of Technology

THE GREAT STORY OF THE WORLD

Near the dawn of known time, before any hint of an Earth or a Sun, the pre-stellar universe was a realm of dark matter and incandescent gas that expanded, cooled, and faded to darkness as space itself stretched. There were ripples in that sea of matter, slight concentrations of mass that drew more mass together, collapsing to form the seeds of galaxies.

The physics of gravitation—and of light, heat, and fluid dynamics—describes these collapses along with the lesser collapses within them that gave birth to a generation of stars that brought light into a time of universal darkness. Physics describes the thermonuclear reactions that lit these stars, fusing hydrogen into heavier elements, and physics describes how they came to explode as supernovas, spreading the stardust that later collapsed to form planets around stars like the Sun.

Then eons later, on a planet simmered in sunlight for a third of known time, a species emerged that had hands, eyes, and minds, and with these, a grasp of the technologies of wood and hide, of bone and stone.

From these beginnings, our ancestors learned to make better tools, and with these in hand, yet better ones. Though slow at its start, this cycle of progress accelerated until, within the span of a few hundred generations, humans had learned the rudiments of blacksmithing. The efforts of mere tens of generations led to machines of steel, and within just a few more, to the tools that enabled human minds and hands to build the instruments that, within the span of just my grandfather's lifetime, brought the first knowledge of the physical laws that govern the hardness of stone, the light of the stars, and the form of the universe.

And with new tools in our hands and minds filled with new knowledge, we learned to hurl finely wrought telescopes far past the Moon, to shield them from the Sun, to chill them colder than space, and through their wide eyes take the measure of faint thermal patterns in ancient light from the incandescent dawn of known time. These faint thermal patterns reveal the slight concentrations of mass that set matter in motion on the path to form stars.

Today, most knowledge of the physical universe has been gained in the span of a lifetime, while the generations of technology, once measured in centuries, now pass more swiftly than our own.

THE PHYSICS OF TECHNOLOGY

What we've learned is immense and yet we've barely begun to understand what it can tell us about our world and what it can become. The same physical knowledge that reaches deep into space and time applies equally to Earth and human technology, to every chip of flint and nanoscale transistor, to every tool and device in our homes, skies, and factories. Indeed, physical law describes far more than this. It defines an abstract part of the shape of the world: the physical potential of technology.

The physical potential of technology is a timeless aspect of physical law, a latent structure older than stars and more solid than stone. Physical law governs the universe and everything in it; it governs every device that does, will, or could exist, and thereby governs the ultimate potential of technology as a whole. Regardless of whether we grasp it or not, the potential of technology is the same here and now as it will be in our future, and the same as it was a billion years in the past.

With the aid of exploratory engineering design and analysis, physical principles can be used to survey a portion of the physical potential of technology, and within a limited range, advanced APM-based technologies are part of the view. To understand the scope of exploratory engineering, however, it is important to recognize the enormous range of questions that these studies can't even begin to answer.

Most views of the future of technology start with the facts and trends of a partly known present then proceed to speculate about near-term discoveries, inventions, investments, costs, markets, and competing technologies, all contingent upon the cascading events of an unknown future. The results of these speculations can be useful, yet their contingency on unknown events yet to come makes them profoundly, irreducibly uncertain.

To study the fundamental physical potential of technology is to ask a different question. The answer can yield information that pertains to our future, but this answer will depend neither on speculations about future events, nor on facts about the past or present state of the world. To study these questions means exploring, not the time-bound consequences of human actions, but the timeless implications of known physical law.

I sometimes picture the realm of potential technologies as a timeless landscape contoured and bounded by physical law. In this picture, potential technologies are places whose distances are measured by accessibility and performance. This landscape holds every potential technology, whether in use, in view, or unimagined. One might call it "possibility space."

To venture outward across this terrain, in concrete physical reality, requires increasing technological capabilities, while the limits of the landscape—the far shores of the continent—correspond to the ultimate limits of technology considered as a whole.

Within this broad landscape of potential technologies we now occupy a settled region that spans the technologies and products of our material civilization. This is a locale marked by rolled steel and automobiles, rocket engines and satellites, by nanoscale transistors and laptop computers. Expanding the settled region is an ongoing process of increasing the performance of technologies—stronger materials, more efficient engines, faster computers.

Just beyond the frontier of today's settled region is a land that contains next year's computers and photovoltaics; a few steps beyond this neighboring land one finds technologies that exist only as engineering concepts, and yet further beyond are regions that hold potential machines and devices of yet-unknown designs—computers far beyond the limits of silicon-chip performance and photovoltaics that approach the limits of sunlight conversion efficiency.

Today's settled region amounts to no more than a patch on the margin of a vast continent of technological potential. The shores of this continent—the limits of the landscape—mark the timeless bounds set by physical law. Some bounds have been reached; others still lie far in the distance, often too far to see.

Physical law is the bedrock beneath the landscape of technological potential, and knowledge of physical law is the basis for exploring the terrain. This exploration can map paths toward APM and can survey at least part of its potential.

The idea of using scientific knowledge to explore potential technologies raises a question: "What counts as *trustworthy* scientific knowledge?" and turning to basic, APM-enabling technologies, a second question: "Is current knowledge enough for the purpose at hand?"

TEXTBOOK SCIENCE TESTED DAILY

When I speak of "scientific knowledge" of physical law (and of knowledge of empirical physical facts), I mean reliable, quantitative knowledge. And here I am using the word "reliable" in a literal sense, to describe the kind of knowledge that scientists and engineers rely on, knowledge that they use as routinely as they use addition, multiplication, and calculus. Textbooks teach this reliable, quantitative knowledge and it's tested in practice every day.

Engineers use this knowledge in designing engines and aircraft, transistors and supercomputers. When a product fails a test, they seldom look for errors in the textbooks—they look instead for errors in the design, analysis, or manufacturing, or for errors in the test itself.

Scientists use this same textbook-quality knowledge as a basis for seeking new knowledge. Designing experiments is a form of engineering, and wherever it's appropriate the assumption that force = mass × acceleration enters into both the design of an experiment and the analysis of its results (and this is good, standard practice, despite Einstein's corrections to Newton's Laws).

Textbook science like this is seldom explicitly questioned, yet through use it's tested every day, and stringently. The vibration of a nanoscale rod, of a bell, and of the Earth are described using the same equations of motion, and every precision measurement tests both what is measured and the assumptions implicit in designing the test. Stubborn discrepancies would draw attention, as they did in the past, and to discover a deep enough discrepancy would bring awards and enduring fame.

Nonetheless, almost all the tested, textbook physics in practical use is known to be at least a little bit wrong.

For example, the breakthrough physics of 1687—Newton's mechanics and law of gravity—doesn't fully accord with reality. Experiments agree instead with the different physics of 1905 and 1915, Einstein's mechanics and his description of gravitation as mass curving space and curved space guiding motion.

Nevertheless, Newtonian mechanics still rules (most) textbooks and (most) practical calculations in fields throughout (most of) science and engineering. Why is this incorrect theory (almost) pervasive? Only because it's expressed by simpler math and because the discrepancies between Newtonian mechanics and reality are, in (most) common circumstances, measurable only by extraordinarily precise instruments. In practice, of course, (almost) no one in science or engineering (in most typical contexts) clutters descriptions with qualifiers (at least of this sort). There's no need to. Newtonian mechanics is simply used without comment.

The deepest understanding of physical law—its most universal and precise formulation (setting gravity aside)—rests on quantum field theory together with a collection of elegant mathematical symmetries that are connected in awkward, asymmetrical ways and filled out with about twenty unexplained parameters derived from experiment. However, the mathematical bridge that links this deepest physical model to everyday physical predictions charges an impossibly steep toll in computational complexity and cost, and so scientists and engineers instead use shallower models of physical law and avoid using levels deeper than they need. When simpler calculations yield results that are, within the limits of measurement, identical to deeper theory, the differences don't matter and the simpler calculations win.

Scientists and engineers aren't alone in using approximations. Architects, for example, make slightly false assumptions about geometry. If a building were built with genuinely vertical walls, its width would grow with its height, and if built with walls exactly on north-south lines, its rooms would taper in width toward the nearest pole. For buildings the size of Colorado, these facts would matter, but they're ignored in practice and disappear into the margin of error in construction and measurement. Just as mechanical engineers calculate as if Newton were right, architects calculate as if the Earth was flat, and they'd be silly if they didn't.

The day-to-day practice of science and engineering reveals the power of modern physical knowledge. So far as fundamental physical law is concerned, the supply of knowledge far exceeds the demand. In exploring the potential of atomically precise manufacturing, for example, it turns out that the relevant principles of physical law have been known since the days of the Model T Ford.

Why, then, is physics a vibrant field today, along with chemistry and a host of other fields in the physical sciences? And if new discoveries continue, how much of the timeless landscape of technological potential is actually visible within the boundaries of current knowledge? To get a good grasp of the contours of knowledge requires a closer look at what still isn't known.

UNIVERSAL YET LIMITED PHYSICAL KNOWLEDGE

Well-tested textbook physics seldom merits a headline, because there's nothing new about knowledge that's been used every day for decades. Instead, science news brings us a flood of reports of discoveries (and of well-hyped, speculative hints of discoveries, and exciting would-be-ground-breaking mistakes), and these reports, by their very nature as news, must highlight unknowns. Thus, even when news about science is accurate, it will present a skewed view of the scope of reliable knowledge.

Physics provides the best example of stable, reliable knowledge. In physics at its most fundamental level, the discovery of anything wholly new earns a Nobel Prize. Viewing science from this most fundamental level offers a useful perspective on the more mundane physics that describes molecules and mountains, the physics that describes the contours of the landscape of potential nanotechnologies from here to APM, and far beyond.

To begin at the bottom, the most fundamental level of physics (the level mentioned above) describes a set of elementary particles and their interactions, which can be summarized by a single terse page of mathematics. A tensor equation that fills half a line describes the whole of General Relativity, Einstein's description of gravitation in terms of curved space and time. Aside from gravitation, the equations that fill the rest of the page—the mathematical mosaic that defines the Standard Model—describe the behavior of every interaction and particle yet observed, with Special Relativity inherent in their very structure. These equations have been tested to the utmost limits of human ingenuity and budgets, to the extremes of experimental conditions and measurement accuracy.*

In 2012, physicists at CERN reported the discovery of the expected Standard Model Higgs boson (or a particle as yet indistinguishable from it), an achievement that required billions of dollars of equipment and enough power to light a small city. With the Higgs boson, physicists observed the final particle predicted by the Standard Model, a crowning yet frustrating achievement.

Physicists had hoped that they would find something else, a terrestrial clue to physics beyond the Standard Model. There must be something beyond. Galactic dynamics shows the effects of mass in the form (it

---

* There's an engineering principle here, by the way: *Phenomena that can't be observed—despite the most strenuous efforts—are extremely unlikely to either impede or enable accessible technologies.*

seems) of unobserved dark-matter particles, and the expansion of the cosmos itself is being driven ever faster by an unexplained dark-energy field. Beyond this, the equations themselves call out for amendment or replacement. Gravitation doesn't fit with the rest (but only deep down, in as-yet unobservable ways), and even the compatible parts of the Standard Model have a patchwork quality, where at the bottom, physicists hope to find a more unified fabric.

Nonetheless, when taken together with General Relativity, the Standard Model describes all the deep physics that matters to life. Indeed, for essentially every practical purpose, much less is sufficient. Caltech physicist Sean Carroll states this point well:

> Over the last four hundred or so years, human beings have achieved something truly amazing: we understand the basic rules governing the operation of the world around us. Everything we see in our everyday lives is simply a combination of three particles—protons, neutrons, and electrons—interacting through three forces—gravity, electromagnetism, and the strong nuclear force. That is it; there are no other forms of matter needed to describe what we see, and no other forces that affect how they interact in any noticeable way. And we know what those interactions are, and how they work. . . . As far as our immediate world is concerned, we know what the rules are.

Although Carroll outlines a fragment of physical knowledge that omits all but a few known physical particles, even this small fragment includes all the physics that affects the behavior of machines, molecules, and APM systems.

The Standard Model stands at a distant frontier of physics, but quite different frontiers are much closer. Some of these are also called "fundamental," and if viewed from the right angle, they earn the name.

I'm using the qualifier "fundamental" in the narrowest sense, to refer to knowledge of matter at the level of its seemingly ultimate elementary particles. In practice, however, almost the whole of physics pursues challenges elsewhere (in fields such as geophysics, astrophysics, and fluid dynamics), while some fields study phenomena that, in a mathematical sense, amount to new particles. These fields of physics study emergent patterns of (quasi)particles in crystalline matter, including work that can require the full mathematical apparatus of quantum field theory. Nobel Prize–winning discoveries in this domain include high-temperature superconductors and more subtle and exotic phenomena like the fractional quantum Hall effect, in which (quasi)particles constrained to move in only two dimensions interact in ways that are topologically impossible in a three-dimensional space.

These phenomena (as is typical of exotic quantum effects) are challenging to observe; many of them occur only in unusual forms of crystalline matter and at temperatures within a fraction of a degree of absolute zero (far colder than a plutonian night).

The engineering principle noted above—in essence, that what can't be observed can't be applied—has a parallel here: *Exotic effects that are hard to discover or measure will almost certainly be easy to avoid or ignore.*

There's another, contrasting principle, however: *Exotic effects that can be discovered and measured can sometimes be exploited for practical purposes.*

For example, although they were discovered a quarter century ago, high-temperature superconductors can still be counted as both exotic and useful. These synthetic, layered cuprate materials have been used to build high-field electromagnets, yet exactly why they act as superconductors (the specific nature of the underlying, collective, correlated-electron effects) remains controversial. For another example that stems from collective electron behavior, the fractional quantum Hall effect (observed only at extremely low temperatures) is now being studied as a potential basis for a robust, topological form of quantum computing.

Thus, discoveries of phenomena in unusual forms of matter won't confront engineers with constraints that apply to less exotic circumstances, but they can (and have) revealed potential capabilities that hadn't been known, new and unexpected regions of the timeless landscape of potential technologies. We can expect to see more.

FROM QUANTUM MECHANICS
TO SPRINGS AND ELASTIC SPHERES

To reach the physics in actual use within the molecular sciences (and in materials science, and engineering) takes us far from the deep fundamentals. The practical level isn't the Standard Model, developed to describe events in high-energy particle collisions (which took shape through multiple Nobel Prize–winning advances in recent decades), nor the exquisitely accurate quantum electrodynamics of the late 1940s (which earned Nobel Prizes for Feynman, Schwinger, and Tomonaga), nor even the relativistic quantum mechanics of the late 1920s (which earned a Nobel Prize for Dirac). Instead, most research—if it applies quantum mechanics at all—relies on computational approximations to a set of mathematical approximations to an approximate *model* of quantum mechanics that dates from the mid-1920s, the theory that earned a Nobel Prize for Schrödinger and Heisenberg.

Researchers use approximations to Schrödinger's equation to calculate various properties of matter: the density of iron, the elastic moduli of silicon carbide, the binding energy of a covalent bond. Through mathematical methods and algorithms refined year after year, using ever more powerful computers, Schrödinger's quantum mechanics has enabled calculations that achieve "chemical accuracy," that is, the ability to calculate energy differences smaller than the energy of room-temperature thermal vibrations.

For both mathematical and computational reasons, however, quantum methods become increasingly impractical when the number of atoms grows large. In practice, the larger the molecular system, the greater the approximations required and the less accurate the answer, eventually to the point of becoming useless. The solution to this problem, of course, is to shift to yet another kind of approximation—one that can give accurate-enough answers for larger-scale systems, provided one asks suitable questions.

In computational chemistry, molecular dynamics methods set twentieth-century quantum mechanics aside in favor of the physics of 1687, applying Newton's equation of motion, $F = ma$, to calculate how atoms move. Through approximate descriptions of interatomic forces, these methods compute the trajectories of atoms as if they were planets following trajectories determined by the interplanetary forces of gravity.

Good descriptions of interatomic forces are the key to accurate results. Quantum calculations can provide the parameters used to construct these approximate descriptions, and in a final step away from the heights of physical theory, the required force parameters have long been derived, not from calculations, but empirically, from laboratory data.

Whatever their sources, the parameters are slotted into equations that are then used to calculate forces from moment to moment, determining the dynamics as atoms move. The force-field equations (just approximations, of course) typically model atoms as elastic spheres that can attract or repel one another and represent bonds between atoms as springs (more or less). These standard molecular-dynamics models were the basis for my earlier description of the look and feel of the nanoscale world.

How good are these approximations to molecular physics? One can quantify their accuracy and discrepancies in quantitative terms, but in practical terms there can be no simple, universal answer, because different practical problems raise different questions.

For example, how a protein chain folds to form a compact, functional object depends on an often delicate balance of interaction energies among thousands of atoms. At present—for natural proteins, i.e., proteins that have not been engineered for high stability—computational methods have only recently gained useful predictive value. Too many configurations are too nearly alike, and subtle entropic differences can tip the balance.

The analogous structure-prediction question is easy to answer when considering nanoscale structures knit together by dense networks of covalent bonds. Structures like these have no flexible chains to fold or unfold, hence no delicate balance of interaction energies or subtle entropic differences among alternative states. Here, the vexing questions raised by protein structures simply don't arise.

Structures of this rigid, predictable sort will become readily available further along the road as we approach APM-level fabrication capabilities. If one asks functional, engineering questions about these kinds of structures, all reasonable computational methods will give the same answer, provided the designs are robust.

Thus, in comparing problems in engineering and science, one often encounters quite different questions and objects of study, and what counts as a good-enough calculation typically differs as well—and indeed can differ enormously. Here, engineers have a luxury. Within broad limits they can choose both their systems and their questions to fit within the bounds of knowledge and calculation. This freedom of choice can help engineers design artifacts that will perform as expected—provided, of course, that the unknown is somehow held at bay.

CONFRONTING THE UNKNOWN AND THE UNPREDICTABLE

If physics is like chess, as Sean Carroll suggests, then it's a game played on a board that's largely hidden from sight, vast in extent, and occupied by countless pieces, all moving. Understanding the intricate patterns of human biology for example—the fabric of substance and causality that links eyes, guts, skin, and brain—can't be simply a matter of physics, because fundamental physics describes only the microscale rules, not the complexity found in the world.

Whether measured by headcount or budget, most of science studies complex things like bacteria, human beings, geology, and climate. Although *Science* and *Nature* publish research that spans the whole of science, or nearly enough, one only occasionally sees papers on fundamental physics. Instead, alongside studies of artificial systems, one encounters a torrent of discoveries about the natural world, papers on molecular and cell biology, papers on the life sciences at higher levels of organization (ecology, brain science, sociology), and yet more papers describing discoveries in geology, climatology, astrophysics, and the structure of the cosmos. All these fields can be seen as emerging from physics, yet their rich complexity stems from compounding and interwoven causal chains that reach back to the origin of the universe.

No physical calculation, no tabulated data can tell us about parts of nature that haven't yet been discovered. What's more, explorations to date have revealed systems of such staggering, layered complexity (the human brain, for example) that efforts to understand them barely touch on physics, and instead reach far and deep into questions of structure and function.

The world's messy complexity often gives engineers reason to shield their creations inside a box of some sort, like a computer or printer in a case, or an engine under the hood of a car. A box provides insulation against the complex, uncontrolled, and unpredictable world of nature (and children).

The external world still causes problems at boundaries, of course. Consider a ship, or a pacemaker. Ship hulls keep out the sea, but the sea attaches barnacles; engineers hide a pacemaker's electronics inside a shell, but the body's immune system and tissue remodeling make medical implant design far more than an exercise in physics. This kind of problem doesn't arise for an APM system protected by a box on a table, which is sheltered in turn inside the box-like walls of a house.

Engineers can solve many problems and simplify others by designing systems shielded by barriers that hold an unpredictable world at bay. In effect, boxes make physics more predictive and, by the same token, thinking in terms of devices sheltered in boxes can open longer sightlines across the landscape of technological potential. In my work, for example, an early step in analyzing APM systems was to explore ways of keeping interior working spaces clean, and hence simple.

Note that designed-in complexity poses a different and more tractable kind of problem than problems of the sort that scientists study. Nature confronts us with complexity of wildly differing kinds and cares nothing for our ability to understand any of it. Technology, by contrast, embodies understanding from its very inception, and the complexity of human-made artifacts can be carefully structured for human comprehension, sometimes with substantial success.

Nonetheless, simple systems can behave in ways beyond the reach of predictive calculation. This is true even in classical physics.

Shooting a pool ball straight into a pocket poses no challenge at all to someone with just slightly more skill than mine and a simple bank shot isn't too difficult. With luck, a cue ball could drive a ball to strike another ball that drives yet another into a distant pocket, but at every step impacts between curved surfaces amplify the effect of small offsets, and in a chain of impacts like this the outcome soon becomes no more than a matter of chance—offsets grow exponentially with each collision. Even with perfect spheres, perfectly elastic, on a frictionless surface, mere thermal energy would soon randomize paths (after 10 impacts or so), just as it does when atoms collide.

Many systems amplify small differences this way, and chaotic, turbulent flow provides a good example. Downstream turbulence is sensitive to the smallest upstream changes, which is why the flap of a butterfly's wing, or the wave of your hand, will change the number and track of the storms in every future hurricane season.

Engineers, however, can constrain and master this sort of unpredictability. A pipe carrying turbulent water is unpredictable inside (despite being like a shielded box), yet can deliver water reliably through a faucet downstream. The details of this turbulent flow are beyond prediction, yet everything about the flow is bounded in magnitude, and in a robust engineering design the unpredictable details won't matter.

Likewise, in APM systems, thermal fluctuations are unpredictable in detail, but bounded in magnitude, and a discipline of robust design once again can give satisfactory results.*

Uncertainty has another facet, however, because even potentially predictable outcomes may be beyond prediction in practice. In the molecular sciences, crystal structures provide an example.

Prepare a sample of a new, pure molecular substance, dissolve it, and let the solvent evaporate. The odds are good that the substance will crystallize (a commonplace example of self-assembly), and if it does, you can perhaps persuade a crystallographer to observe how X-rays diffract from planes of aligned atoms in the crystal's interior and from the data infer and report how the molecules pack together to form some particular crystalline pattern.

Can this specific pattern be predicted before studying the X-ray diffraction data? In principle, the answer should often be Yes. The molecules will often assemble to form a unique, minimum-energy structure, a pat-

---

* The same is true of the smaller effects of quantum fluctuations: calculable, bounded, and acceptable.

tern that packs molecules closely, while (for example) placing positive and negative charges together and forming as many hydrogen bonds as possible. In practice, however, the answer will often be No—favorable conditions may conflict with one another (better packing leading to worse charge placement, and so on), and the actual structure that nature will find may depend on a razor's-edge difference in energy balance among several alternative forms.

Like protein-fold prediction, crystal-structure prediction remains a great challenge, one that is largely unsolved. In exact parallel with the case of protein fold prediction vs. protein engineering, however, the field of crystal engineering is nonetheless thriving. Engineers design for predictability, and as protein and crystal engineering illustrate, engineers can succeed at their task when a casual look at the science might suggest that they can't.

In conventional engineering, structures are routinely designed to behave predictably in all the ways that matter to outcomes. Structures hold shapes determined not by thermal motion, but by design and directed assembly. Likewise, machines move in particular ways, not randomly like colliding balls or molecules, but following paths that remain within tolerances, constrained tightly enough that small vibrations won't disrupt their intended functions. These characteristics are true of machines used in conventional manufacturing, and equally true of smaller machines that can be built using APM-level technologies.

ENGINEERING WITHIN THE BOUNDS OF LIMITED KNOWLEDGE

Engineers work with knowledge of limited scope and precision in much the same way that they work with motors of limited power and materials of limited strength. Within specific constraints of knowledge, some tasks can be achieved predictably, while others cannot.

Coping with limited knowledge is a necessary part of design and can often be managed. Indeed, engineers designed bridges long before anyone could calculate stresses and strains, which is to say, they learned to succeed without knowledge that seems essential today. In this light, it's worth considering not only the extent and precision of scientific knowledge, but also how far engineering can reach with knowledge that remains incomplete and imperfect.

For example, at the level of molecules and materials—the literal substance of technological systems—empirical studies still dominate knowledge. The range of reliable calculation grows year by year, yet no one calculates the tensile strength of a particular grade of medium-carbon steel. Engineers either read the data from tables or they clamp a sample in the jaws of a strength-testing machine and pull until it breaks. In other words, rather than calculating on the basis of physical law, they ask the physical world directly.

Experience shows that this kind of knowledge supports physical calculations with endless applications. Building on empirical knowledge of the mechanical properties of steel, engineers apply physics-based calculations to design both bridges and cars. Knowing the empirical electronic properties of silicon, engineers apply physics-based calculations to design transistors, circuits, and computers.

Empirical data and calculation likewise join forces in molecular science and engineering. Knowing the structural properties of particular configurations of atoms and bonds enables quantitative predictions of limited scope, yet applicable in endless circumstances. The same is true of chemical processes that break or make particular configurations of bonds to yield an endless variety of molecular structures.

Limited scientific knowledge may suffice for one purpose but not for another, and the difference depends on what questions it answers. In particular, when scientific knowledge is to be used in engineering design, what counts as enough scientific knowledge is itself an engineering question, one that by nature can be addressed only in the context of design and analysis.

Empirical knowledge embodies physical law as surely as any calculation in physics. If applied with caution—respecting its limits—empirical knowledge can join forces with calculation, not just in contemporary engineering, but in exploring the landscape of potential technologies.

To understand this exploratory endeavor and what it can tell us about human prospects, it will be crucial to understand more deeply why the questions asked by science and engineering are *fundamentally* different. One central reason is this: Scientists focus on what's not yet discovered and look toward an endless frontier of unknowns, while engineers focus on what has been well established and look toward textbooks, tabulated data, product specifications, and established engineering practice. In short, scientists seek the unknown, while engineers avoid it.

Further, when unknowns can't be avoided, engineers can often render them harmless by wrapping them in a cushion. In designing devices, engineers accommodate imprecise knowledge in the same way that they accommodate imprecise calculations, flawed fabrication, and the likelihood of unexpected events when a product is used. They pad their designs with a margin of safety.

The reason that aircraft seldom fall from the sky with a broken wing isn't that anyone has perfect knowledge of dislocation dynamics and high-cycle fatigue in dispersion-hardened aluminum, nor because of perfect design calculations, nor because of perfection of any other kind. Instead, the reason that wings remain intact is that engineers apply conservative design, specifying structures that will survive even unlikely events, taking account of expected flaws in high-quality components, crack growth in aluminum under high-cycle fatigue, and known inaccuracies in the design calculations themselves. This design discipline provides safety margins, and safety margins explain why disasters are rare.

Limited knowledge thus calls for conservative design, and engineers pay for this—through costs in production, costs in performance—and by paying these costs, they buy confidence that what they know, calculate, and build will add up to something that works.

In assessing the landscape of potential technologies, conservative design is an essential tool. Just as conservative design can make products strong and robust it can do the same for the results of engineering analysis. What's more, when the analysis itself is the product, physical costs of production and performance scarcely apply.

WHERE APM STANDS IN THE LANDSCAPE OF TECHNOLOGY

Human knowledge of fundamental physical law embraces everything we can see or make on Earth. To the extent that predictions can be extracted from its mathematics, known physical law describes reality to the limits of measurement.

Long-established principles of physical law, applied in conjunction with empirical knowledge, can predict the behavior of physical systems across a wide range of physical circumstances, often with extraordinary accuracy. This knowledge supports today's engineering design and production, and—crucially—it can support the design and analysis of systems that are beyond the reach of today's manufacturing capabilities.

Seen in this context, APM systems and their potential products have several convenient characteristics. APM-level fabrication, by its very nature, enables the production of extraordinarily strong materials and high-performance devices. Using extraordinarily high-performance components as building blocks relaxes many constraints, enabling the design of systems far beyond today's state of the art that are nonetheless based on conservative design choices in every respect. Better yet, having a wide range of conservative options often provides ample room for crafting system architectures that avoid the unknown, taking care to build only on firm ground.

Through the dual lenses of science and engineering, we can see landmarks at a distance across the landscape of technological potential. Not in all directions, or in full detail, yet what stands in plain sight includes APM and the potential for a future of radical abundance.

What stands in plain sight through these dual lenses, however, has in practice been hard to see. The lenses, it seems, have been pointing in different directions, impairing our vision.

CHAPTER 8

The Clashing Concerns
of Engineering and Science

OF ELEPHANTS AND AUTOMOBILITY:
TWIN PARABLES OF SCIENCE AND TECHNOLOGY

The story of the blind men and the elephant originated in South Asia uncounted centuries ago and no one now knows its source, though the tale has long been part of the Buddhist, Hindu, Jain, and Sufi traditions. It took root in the English-speaking world through a poem penned over a century ago:

> *It was six men of Hindustan*
> *To learning much inclined,*
> *Who went to see the Elephant*
> *(Though all of them were blind),*
> *That each by observation*
> *Might satisfy his mind . . .*
>
> —John Godfrey Saxe (1816–1887)

In the story as the Jains tell it,

The blind man who feels a leg says the elephant is like a pillar; the one who feels the tail says the elephant is like a rope; the one who feels the trunk says the elephant is like a tree branch; the one who feels the ear says the elephant is like a hand fan; the one who feels the belly says the elephant is like a wall; and the one who feels the tusk says the elephant is like a solid pipe.

The blind investigators fall into dispute until a wise man enlightens them, explaining that each has grasped one part of a many-sided truth.

The start of this tale can serve as a metaphor for science, but not its end. In science, there can be no wise answer from outside the system, nor would scientists be satisfied with such a fragmented view. Instead, the process works something like this:

The Blind Scientists and the Elephant: A Parable of Successful Science

Once upon a time, a group of scientists in the burgeoning field of megabeastology came together to form a new specialty: elephantology. As it happens, they too were blind, and like the blind men in the original story, they studied different parts of an elephant, perceiving just one part at a time.

As senior scientists, they had risen to the status of independent investigators and could choose their research directions. Some measured the texture and curvature of different patches of skin, finding very different results, and spectral analysis revealed correlations between color and roughness. Some studied fluid influx, others studied efflux, and they compiled statistical time series and noted a diurnal periodicity. The leading researchers published their studies in the prestigious *International Journal of Elephantology*, and in its rivals, the *Journal of Pachyderm Science* and *Megabeast Letters*.

Some of the earliest research results enabled scientists to reinterpret the literature on anomalous tree mobility, generating shock-waves and disbelief in the dendrology community. Within elephantology, meanwhile, clashing observations created turmoil. Regarding the puzzle of elephant geometry, the cylindrical, spherical, and rope-theory models all had their advocates.

Over the years, however, with improving instruments and continued funding, our scientists extended their research and found that they were exploring a single surface, but from different directions. Clashing models were abandoned or reconciled, piece by piece, ultimately resulting in two Nobel Prizes and the well-known Standard Elephant Model, a triumph of science.

THESE INDEPENDENT investigators studied different parts of the elephant using different methods, yet in the end, with little coordination, they produced a coherent result. What gave independent research its coherence is simple, of course. Their observations of different parts all fit together because the parts themselves fit together. They all studied the same elephant, and it existed before they began.

The Independent Automobilists: A Parable of Failed Engineering

Once upon a time, scientists drawn from a wide range of fields banded together to pursue a grand, visionary challenge: to develop the world's first automotive machine for passenger transport. Our researchers proposed links between this vision and their scientific specialties, and agencies funded their work. Papers followed, funding grew, and the pursuit of automobility spawned research centers, national programs, and journals with names like *Automotive Letters* and the *Journal of Autoscience and Autotechnology*. Within a decade, Congress had launched a National Autotechnology Initiative.

While promising a technological result, however, the field of automobility research retained the structure of science. Curiosity-driven investigators chose their own objectives and shared their knowledge in journals, and each group had its own ideas of what might contribute to progress.

Hazy concepts of automotive machines took shape and proliferated. Automobilists (as the researchers came to be known) saw a need for some sort of motive power, and so specialists in different fields came up with a wide range of different ideas. Some studied materials for motors powered by coiled springs, while others pursued wind power, or piston engines, or experimented with the use of exercise wheels to harness the power of hamsters.*

The idea of a passenger enclosure spurred studies of fabric for tents, wood for boxes, steel sheets for canisters, and adobe bricks for walls that might somehow be made to move. Meanwhile, the idea of mobility itself spurred studies of robotic legs, greased roads, wheels, and rollers. This early automobility research succeeded (it seemed), because it did produce fabrics, boxes, rollers, hamster wheels, and road grease.

Reports of progress came daily in every field, and for a decade it seemed that humankind was on the verge of the heralded age of automobility, yet year after year no one delivered a vehicle—not even a prototype.

Eventually, however, a loose federation of automobilists began to focus on steel boxes mounted on low-friction wheels and on burning fuels to drive pistons in engines of various kinds, and even on ways of making engines turn wheels. Unfortunately, research on twisting wheels sideways or increasing their friction never became fashionable, and so steering and braking remained unsolved problems. The second-order problem of enabling a car to turn as it moves forward—the problem of coupling engine torque to drive both the left and right wheels as they roll different distances—went wholly unrecognized. (This is the problem solved by differential gears, an old but not obvious invention.)

The vision of automobility drew brilliant minds and billions of dollars and advanced a wide range of useful technologies, yet the research fell far short of the vision. Automobilists never delivered a vehicle, or even a working gasoline engine. As years stretched into decades, the promise of automobility receded into a vague and distant future. To some, automobility seemed to be an impossible dream, a vision forever beyond human reach.

---

* In reality, no serious automotive proposal could be this absurd, yet in current research under the nanotechnology umbrella, invisibly small things that are called "motors" only rarely resemble a useful device.

It came as a shock when a group of closely coordinated engineering teams refined the vision into a design, filled the technology gaps, put the pieces together, and delivered the first working automobile.

====

HOW COULD WELL-FUNDED automobility research fail so completely, and for so many years?

Simply because the field and its funders had, in fact, organized a research program modeled on science, to the exclusion of engineering development. Independent automobilists, no matter how brilliant, could make progress forever, yet never fill all the technology gaps. Worse yet, even if all the gaps had been filled, a host of independent researchers could never produce the thousands of different, complementary parts that fit together to make a functional automobile. In other words, regardless of the researchers' brilliance, a program without engineering leadership could never produce a machine like a Model T Ford.

Independent elephantologists could deliver coherent results because elephantology began with elephants—concrete physical systems with a natural, preexisting unity. Automobility, by contrast, began with no automobile, no existing, external object of study to bring coherence. Coherent development requires coherent designs, but designs can't be found by studying objects in nature; designs require the work of human minds, and physical objects come after.

The Genome and the Moon

Turning to history, two great projects of the twentieth century illustrate the contrast between engineering and science. The first reached for the Moon, the second read the human genome. Both projects mobilized large-scale efforts and both succeeded, but they pursued their goals in totally different ways.

In 1961, the Apollo project began with an abstract design, a quantitative description (though not fully detailed) of a system based on liquid fuel rockets, each a complex artifact in itself and part of a more complex system. In the first-stage rockets, fuel pipes had to fit fuel tanks and pumps (fitting in every aspect of form and function), while the pumps fit the engines that fit the booster that launched the upper stages, carrying lunar vehicles, astronauts, spacesuits, and boots to an altitude of sixty-five kilometers and a speed near Mach 8, where the second stage engines ignited.

Engineers first developed a system-level design (evolving through second, third, and tenth drafts), then refined their designs far enough to delegate further refinement to engineers who shaped more detailed designs through delegated choices that shaped details within details until the shape and structure of every part had been fully defined. In July, 1969, Neil Armstrong stepped onto the lunar surface, confirming that the engineers had gotten the details right.

Apollo has often been described as a triumph of science, but this is mistaken: In the grand scheme of things, the new science required was routine and peripheral. Apollo can be better described as a triumph of engineering, and while this is inarguable, the engineering was also routine—in detail. The Apollo program, viewed in perspective, can best be described as a breakthrough in engineering *management,* an effort unprecedented by metrics of scale, complexity, and tight, reliable coordination.

Within the field of molecular biology, the Human Genome Project was also unprecedented in terms of scale and coordination, but these were of a different and simpler kind—the scale of its funding and the challenge of persuading independent scientists to cooperate at all. Fortunately, a suitable pattern of organization was almost inherent in the task. Nature groups genes into chromosomes, and the Human Genome Project divided the task of sequencing different chromosomes among different countries and research groups.

In broad outline, that's all that was necessary—funding, commitments, and a division of labor. Innovations in technologies and methods were crucial (better computational tools, laboratory techniques, and methods like shotgun sequencing, for example), but advances like these didn't demand detailed coordination. In the end, the coherence of nature itself ensured a coherent result.

The Human Genome Project, in effect, began with an elephant, while Apollo began with a conceptual system design. One succeeded through almost independent inquiry; the other demanded the tightest project integration that the world had ever seen.

These radically different modes of organization emerged from fields with radically different cultures—different in terms of patterns of work and thought, different in terms of conceptions of problems and how to solve them. If space systems engineers and molecular biologists had been thrust together, each might have regarded the other as a group trained by aliens from Mars.

WHY SCIENCE AND ENGINEERING
FACE OPPOSITE DIRECTIONS

At the deepest, epistemic level, scientific inquiry and engineering design face opposite directions. While both inquiry and design link patterns in human minds to patterns in matter, their information flows in opposite directions. Inquiry and design are as different as image data from a camera and instructions sent to a printer, as different as sensory and motor neurons, as different as an eye and a hand.

The essence of science is inquiry; the essence of engineering is design. Scientific inquiry expands the scope of human perception and understanding; engineering design expands the scope of human plans and results.

Inquiry and design are perfectly distinct as concepts, but often interwoven in practice, whether within a field, a research program, a development team, or a single creative mind. Meshing design with inquiry can be as vital as hand-eye coordination. Engineering new instruments enables inquiry, while scientific inquiry can enable design. Chemical engineers investigate chemical systems, testing combinations of reactants, temperature, pressure, and time in search of conditions that maximize product yield; they may undertake inquiries every day, yet in the end their experiments support engineering design and analysis. Conversely, experimental physicists undertake engineering when they develop machines like the Large Hadron Collider. With its tunnels, vacuum systems, superconducting magnets, and ten-thousand-ton particle detectors, this machine demanded engineering design on a grand scale, yet all as part of a program of scientific inquiry.*

But the close, intertwining links between scientific inquiry and engineering design can obscure how deeply they differ. To understand the difficulty, consider how much they share.

Engineers and scientists both work with physical systems described by universal physical laws. Both share the language of mathematics, the methods of algebra, calculus, and differential equations, and they apply these methods to the same laws of motion, forces, and fields. Both likewise share technical vocabularies used to discuss classical mechanics, fluid dynamics, materials, optics and more, and often use these vocabularies to describe the same materials, and sometimes the very same physical objects.

What's more, scientists and engineers frequently collaborate and their roles can even merge. Engineers build on knowledge uncovered by scientists, while scientists use instruments built by engineers. Scientists often design their own instruments while engineers often make measurements that contribute to science, merging both roles in a single mind. It's no wonder that science often gets credit for engineering accomplishments, while engineering is often mistaken for a kind of science.

Turning to the larger-scale texture of research, a scientific project may consist mostly of engineering, if counted by dollars or brain-years, and not by its ultimate purpose, and vice-versa for engineering. Some fields of science are engineering intensive—planetary science, for example, is joined at the hip with space systems engineering. Likewise, some fields of engineering are science intensive. Scientific journals are filled with reports of new ways of making useful things—reports of engineering progress enabled by science-rich research.

---

* When I contrast patterns of thought, values, culture, and institutions in science and engineering, I intend to refer to contrasts in typical patterns in fields that center on inquiry, or center on design. Labels for fields may not characterize their substance, and seldom do so fully. Departments of computer science typically focus on advancing the arts of design, while departments of mechanical engineering often embrace programs devoted to materials science.

In particular, new nanoscale structures often stem from new processes, and both the structures and processes become objects of inquiry before they can become building blocks for systematic design. "Nanoscience" and "nanotechnology" have often been used interchangeably, and this close, practical relationship has been one of the reasons.

The integration of design and inquiry reaches further. A plan to drive to a store can be seen as design, and searching for a set of car keys as a kind of inquiry. In the most fundamental sense, the intellectual activities of inquiry and design work together in every mind as surely as there are sensory neurons and motor neurons, sight and motion, recognition and intention. Indeed, in this sense, the dual processes of design and inquiry can be found in the active and curious mind of a mouse.

In light of all these connections, how deep is the difference between science and engineering? Deep enough to engender structural contrasts that run from bottom to top. The contrasts have roots as deep as epistemology, a contrast in the very direction of information flow between mind and matter. From this root cause, the contrast extends upward and spreads through education, work, intellectual values, and institutional structures.

Engineering and science engage with the same physical world, yet design and inquiry, by nature, turn minds to face opposite directions.

The Bottom-Up Structure of Scientific Inquiry

Scientific inquiry builds knowledge from bottom to top, from the ground of the physical world to the heights of well-tested theories, which is to say, to general, abstract models of how the world works. The resulting structure can be divided into three levels linked by two bridges.

At the ground level, we find physical things of interest to science, things like grasses and grazing herds on the African savannah, galaxies and gas clouds seen across cosmological time, and ordered electronic phases that emerge within a thousandth of a degree of absolute zero. In principle, things of interest to science extend to the totality of the universe and every potential phenomenon.

On the bridge to the level above, physical things become objects of study through human perception, extended by instruments like radio telescopes, magnetometers, and binoculars, yielding results to be recorded and shared, extending human knowledge. Observations bring information across the first bridge, from physical things to the realm of symbols and thought.

At this next level of information flow, scientists build concrete descriptions of what they observe. From ancient photons captured by telescopes, astronomers infer the composition and motion of galaxies; by probing materials with electronic instruments, physicists infer the dynamics of systems of coupled electrons; by using their eyes and telemetry signals, ornithologists study how bar-tailed godwits cross from Alaska to New Zealand, spanning the Pacific in a single non-stop flight. Through data, scientists describe what they seek to explain.

On the bridge to the top level of this sketch of science, concrete descriptions drive the evolution of theories, first by suggesting ideas about how the world works, and then by enabling tests of those ideas through an intellectual form of natural selection. As theories compete for attention and use, the winning traits include simplicity, breadth, and precision, as well as the breadth and precision of observational tests—and how well theory and data agree, of course.

Newtonian mechanics serves as the standard example. Its breadth embraces every mass, force, and motion, while its precision is mathematically exact. This breadth and precision are the source of both its power in practice and its failure as an ultimate theory. Newton's Laws make precise predictions for motions at any speed, enabling precise observations to reveal their flaws.

Thus, in scientific inquiry, knowledge flows from bottom to top:

- Through observation and study, physical systems shape concrete descriptions.
- By suggesting ideas and then testing them, concrete descriptions shape scientific theories.

Figure 1 illustrates this schematic structure of scientific inquiry, contrasting it with the antiparallel structure of engineering design.

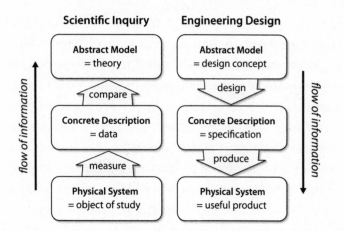

FIGURE 1: The Antiparallel Structures of Scientific Inquiry and Engineering Design

The Top-Down Structure of Engineering Design

In scientific inquiry information flows from matter to mind, but in engineering design information flows from mind to matter:

- Inquiry extracts information through instruments; design applies information through tools.
- Inquiry shapes its descriptions to fit the physical world; design shapes the physical world to fit its descriptions.

At this level, the contrasts are often as concrete as the difference between a microscope in an academic laboratory and a milling machine on a factory floor. At the higher, more abstract levels of science and engineering, the differences are less concrete, yet at least as profound. Here, the contrasts are between designs and theories, intangible yet different products of the mind.

- Scientists seek unique, correct theories, and if several theories seem plausible, all but one must be wrong, while engineers seek options for working designs, and if several options will work, success is assured.

- Scientists seek theories that apply across the widest possible range (the Standard Model applies to everything), while engineers seek concepts well-suited to particular domains (liquid-cooled nozzles for engines in liquid-fueled rockets).
- Scientists seek theories that make precise, hence brittle predictions (like Newton's), while engineers seek designs that provide a robust margin of safety.
- In science a single failed prediction can disprove a theory, no matter how many previous tests it has passed, while in engineering one successful design can validate a concept, no matter how many previous versions have failed.

With differences this stark, it may seem a surprise that scientific inquiry and engineering design are ever confused, yet to judge by both the popular and scientific press, clear understanding seems uncomfortably rare.* Among its many consequences, this confusion undercuts understanding of the basis of knowledge in engineering—for example, knowledge regarding the potential of APM-level technologies.

To understand the essential nature and power of engineering, the best place to stand is the top of the design process, the systems-engineering view that I learned while immersed in the practice and culture of space systems engineering.

THE STRATEGY OF SYSTEMS ENGINEERING

The view from the top of scientific inquiry is widely understood, in a general way. Theorists seek and test precise explanations of observations of the natural world, and successful theories are, in a sense, predetermined by nature. The view from the top of engineering design is radically different and less often discussed. When engineers architect systems, they make abstract choices constrained by natural law, yet not fully specified and in no sense predetermined by nature.

---

* This morning, in covering a planned North Korean satellite launch, the BBC reported that "It is expected to take scientists several days to assemble the three-stage rocket. . . ." I doubt that these scientists will find a journal to report their discoveries. Or are they perhaps engineers supervising technicians? Like many others, the reporter seems oblivious to the difference.

The key to understanding engineering at the systems level—the architectural level—is to understand how abstract engineering choices can be grounded in concrete facts about the physical world. And a key to this, in turn, is to understand how engineers can design systems that are beyond their full comprehension.

Once upon a time, a blacksmith could know all the tools and materials needed to practice his craft. In later times, as watchmakers learned to make intricate machines, a craftsman relied on a supply chain of tools and materials too extensive to know, yet the product itself could be known and understood by a single mind.

Today, smithing has given way to computer-controlled steel mills and machine tools, and the craft of watchmaking has given way to automated factories that make watches with millimeter-scale gears and electronic devices that count the megahertz oscillations of pieces of quartz. A cartwright could understand a cart in detail, but an aerospace engineer can never fully understand the complexity packed into the wings of a modern passenger jet, nor can anyone even half understand the yet more complex factories that make its components.

How can deliberate design yield products more complex than anyone can understand?

In essence, such products are designed through leadership and delegation, like managing a corporation in which no one person can understand the whole, and in particular, no upper-level manager can fully understand the jobs performed by all the people who do the nonmanagerial work. Engineers in the upper levels of the design process, typically working in teams, play a similar role in shaping broad design choices—the general configuration of a vehicle, the required power of an engine—while delegating a range of design tasks that demand more aggregate knowledge than any one person could possibly command. Making, weighing, and refining strategic-level design decisions comprise the substance of systems architecture. A systems-level building designer—an architect in the usual sense—plays the same role.

A building's architect can specify the position and load-bearing capacity of a wall without first defining the size and position of the individual steel girders. Those details can be hidden and delegated. Likewise for wings of kinds that have been built before. A designer can select a particular length, chord, and configuration (and much more) without delving into the complexity of the machines that move the landing flaps: Their structure and function can be described and then assumed, and if the aircraft design moves forward, detailed machine design will be delegated to a team of engineers at a later date.

Thus, the key to designing and managing complexity is to work with design components of a particular kind—components that are complex, yet can be understood and described in a simple way *from the outside*. This top-down, divide-and-conquer strategy enables what would otherwise be impossible. Without it, there would be no computers, no printers, no cars, and no factories to make even the simplest things.

Designing and building systems this way—as compositions of functional subsystems—enables the division of labor and knowledge. This is how engineers describe gaps to be filled and weigh the rewards of filling them with components that fit. This is how systems-level engineering sets in motion a process that can yield a coherent result with no need for an elephant.

Systems-level engineering is a discipline radically different from science, though it must conform to the same physical reality. Systems-level engineering, moreover, can provide the perspective required to make sense of the tangles of science-intensive research, to recognize gaps, to define roles that define criteria and metrics for judging advances, both achieved and proposed.

APM and the paths that lead there present challenges of science-intensive engineering. There's little confusion between science and engineering when the two are loosely linked—when scientists pursue knowledge with no concern for applications and engineers pursue innovation using textbook-grade science. The lines become blurred, however, when the two activities become more closely woven together, as they are in engineering molecular systems. Here, systems-level engineering objectives can sometimes be outlined with confidence, while detailed design and fabrication often require extensive experimental work. In this kind of science-intensive engineering, the sharp conceptual difference between the two facets—inquiry and design—is easily lost in a kind of intellectual haze.

The opportunity costs of this can be immense.

Complementarity, not similarity, is what makes scientific research and engineering development a powerful team, whether loosely linked or seamlessly woven together, according to the characteristics of the knowledge required for the task at hand.

THE CHALLENGE OF NEEDLESSLY CLASHING CONCERNS

APM will emerge from the molecular sciences through macromolecular systems engineering, but this field has been slow to develop. Progress in AP fabrication technologies has been immense, yet unfocused. Today, most macromolecular engineering research serves divergent purposes— applications in medicine and materials science, imitation of biomolecules out of sheer curiosity, and so on. Although this work has great value, something is missing.

Macromolecular engineering to date has barely begun to develop a broad vision of system-level objectives and where they can lead. Without coordination around a shared vision—one that embraces a full kit of required components—there can be no clear idea of what's needed, nor a coordinated effort to build all the pieces. As in the parable of the automobilists, there can be endless advances that fail to converge on a functional whole.

Science pursues answers to questions, but not always the questions that engineering must ask. For a new field of engineering to emerge from a field focused on inquiry, it's important to understand how the different concerns of engineering and science can become confused, and how clarity can both accelerate progress along a chosen path, and help us find paths that lead to greater rewards.

APM, of course, is a prime example.

Seeking Knowledge vs. Applying Knowledge

Because science and engineering face opposite directions, they ask different questions.

Scientific inquiry faces toward the unknown, and this shapes the structure of scientific thought; although scientists apply established knowledge, the purpose of science demands that they look beyond it.

Engineering design, by contrast, shuns the unknown. In their work, engineers seek established knowledge and apply it in hopes of avoiding surprises. In engineering, the fewer experiments, the better.

Inquiry and design call for different patterns of thought, patterns that can clash. In considering the science in the area around an engineering problem, a scientist may see endless unknowns and assume that scarce knowledge will preclude engineering, while an engineer considering the very same problem and body of knowledge may find ample knowledge to do the job. Waiting for anything like comprehensive knowledge can be a mistake. If transistor-based electronics engineering, for example, had awaited a general understanding of the behavior of electrons in solid materials, there would never have been a transistor technology based on aluminum, silicon dioxide, and silicon. As it happens, all that engineers needed was the narrower slice of knowledge behind the silicon-based technologies that launched the digital revolution.

I've repeatedly encountered misunderstandings that spring from this kind of confusion regarding the role of unknowns.

A lecture at Oxford last year touched on the topic of AP nanomechanical bearings, and afterward I spoke with the speaker, an experimentalist with extensive knowledge regarding atomically precise surfaces in ultrahigh vacuum.

As a criticism of published AP-bearing concepts, he remarked that many surfaces spontaneously reorganize or bond to others in unacceptable ways and that how surfaces will behave is often unpredictable. And indeed, a pair of bare silicon (100) surfaces, for example, would reconstruct in vacuum and bond on contact, while many other surfaces simply aren't understood today.

The bearings of engineering interest, however, are of the kind I described in Chapter 5, devices in which predictably stable surfaces glide over one another with a smooth, nearly frictionless motion. To solve a problem, all that an engineer needs are a few good options (one will do), and for bearings in nanomechanical systems, the Zettl laboratory in

Berkeley has already demonstrated a suitable range of moving parts based on nested graphene cylinders. These structures behave as predicted by standard computational models, and these same computational models point to many other potential designs.

When I mentioned these facts to my interlocutor—who knew all the relevant science perfectly well—he paused and replied "Oh . . . yes, of course," or words to that effect. A scientific mode of framing questions had prompted him to raise questions regarding unknowns, while engineering questions with well-known answers are—almost by definition—of little interest to science.

The moral of the story: When considering an engineering problem, beware of letting related unknowns distract attention from well-understood solutions.

Seeking Precision vs. Exploiting Approximations

Limited precision is a special case of limited knowledge, a kind of ignorance that pervades everything, including engineering design.

When faced with imprecise knowledge, a scientist will be inclined to improve it, yet an engineer will routinely accept it. Might predictions be wrong by as much as 10 percent, and for poorly understood reasons? The reasons may pose a difficult scientific puzzle, yet an engineer might see no problem at all. Add a 50 percent margin of safety, and move on. Safety margins are standard parts of design, and imprecise knowledge is but one of many reasons.

—————

IN TALKING WITH the director of a major nanotechnology research center (a leader in his field), I mentioned modeling mechanical nanosystems using standard molecular mechanics methods for computational molecular dynamics. He objected that these methods are "inaccurate." I agreed that they have too little accuracy for some purposes (for example, some aspects of biomolecular modeling), but noted that they have more than enough accuracy for others, and that the systems I'd studied (rigid, stable gears and bearings, for example) were of the latter sort. He paused, looked thoughtful, and we moved on to another subject.

Accuracy can only be judged with respect to a purpose and engineers often can choose to ask questions for which models give good-enough answers.

The moral of the story: Beware of mistaking the precise knowledge that scientists naturally seek for the reliable knowledge that engineers actually need.

Confronting Natural Phenomena vs. Designing Reliable Products

Nature presents puzzles that can thwart understanding for decades, and perhaps sometimes forever. Predicting the weather, predicting the folding of membrane proteins, predicting how particular molecules will fit together to form a crystal—all of these problems are long-standing areas of research that have achieved substantial but only partial success. In each of these cases, the unpredictable objects of study result from a spontaneous process—evolution, crystallization, atmospheric dynamics—and none has the essential features of engineering design.

What leads to system-level predictability?

- Well-understood parts with predictable local interactions, whether predictability stems from calculation or testing
- Design margins and controlled system dynamics to limit the effects of imprecision and variable conditions
- Modular organization, to facilitate calculation and testing and to insulate subsystems from one another and the external world

In science, by contrast, many systems that are objects of study lack many (or all) of these characteristics.

- The components and interactions of the system in question may be poorly understood, or unknown.

- The system may lack anything like design margins or control, and may even be wildly unstable.
- The system's components may be far from modular, linked by a tangle of causal connections both inside and outside the system.

Indeed, insulating a system from the external world is almost always a first step in achieving predictable, reliable behavior; an engineer wouldn't want to install a bare circuit board in a beehive.

———

A SCIENTIST WROTE an article about nanomachines of the general sort I've described, but he suggested that they couldn't be used in a biological environment because biomolecules would gum up gears and other moving parts. The answer, of course, is to keep gears in a gearbox, and to place all the critical moving parts inside a sealed shell. The idea of putting machinery in a box may seem obvious—why might a scientist overlook it?

As it happens, the article's author has focused his attention on biomolecular machine systems that do indeed work in biological fluids and are adapted to this messy environment. Soft, biomolecular machines in biological fluids raise important questions for science, but the answers often have no importance for a rigid machine in a box.

The moral of the story: When judging engineering concepts, beware of assuming that familiar concerns will cause problems in systems designed to avoid them.

Seeking Unique Answers vs. Seeking Multiple Options

Expanding the range of possibilities plays opposite roles in inquiry and design.

If elephantologists have three viable hypotheses about an animal's ancestry, at least two hypotheses must be wrong. Discovering yet another possible line of descent creates more uncertainty, not less—now three must be wrong. In science, alternatives represent ignorance.

If automobile engineers have three viable designs for a car's suspension, all three designs will presumably work. Finding yet another design reduces overall risk and increases the likelihood that at least one of the designs will be excellent. In engineering, alternatives represent options.

Not knowing which scientific hypothesis is true isn't at all like having a choice of engineering solutions. Once again, what may seem like similar questions in science and engineering are more nearly opposite.

———

I FIND THAT KNOWLEDGE of options is sometimes mistaken for ignorance of facts. Remarkably, in engineering, even *scientific uncertainty* can contribute to knowledge, because uncertainty about scientific facts can suggest engineering options.

For example, years ago, when the mechanism of the bacterial flagellar motor wasn't yet understood, molecular biologists had developed and explored multiple hypotheses. Only one could be valid as a description of nature, yet each then-viable hypothesis suggested a way of engineering molecular motors. Thus, problematic, competing scientific hypotheses can point toward desirable, competing engineering options—indeed, the more difficult it is to reject a potential explanation for how something behaves, the more likely it is that a working device can be built on the same principle.

The moral of the story: Beware of mistaking the power of multiple engineering options for the problem of competing scientific hypotheses.

Testing Theories vs. Designing Systems

Scientific theories and engineering designs contrast profoundly at the level of logic.

In *The Logic of Scientific Discovery*, Karl Popper observed that universal theories can, in principle, be falsified by one decisive experiment, yet cannot be proved by even a million concordant experiments. (After centuries of success, Newton proved to be wrong.)

In logical terms, a universal physical theory corresponds to a universally quantified statement: *"For all* potential physical systems . . . ,"* while an engineering design corresponds to an existentially quantified statement: *"There exists* a potential physical system. . . ."* One counterexample can disprove the first, while one positive example can prove the second.

In this deepest, logical sense, the questions implied by design and inquiry are opposite (in mathematical terms, they are duals).

———

I'VE ENCOUNTERED this kind of confusion again and again when scientists give only superficial attention to questions of advanced nanotechnology. The faulty reasoning, if stated explicitly, is hard to fathom: "I can think of *something* that won't work, therefore I doubt that *anything* can work." Several of the cautionary stories above also illustrate this error:

"*Some* surfaces would bond together on contact, therefore . . ."
"*Some* systems are sensitive to small inaccuracies, therefore . . ."
"*Some* kinds of machines would fail when exposed to gunk . . ."

True statements like these do not speak against well-chosen surfaces serving as bearings, or well-designed systems being robust, or machines working because they don't let gunk get inside. At best, they draw attention to constraints.

Critiques that begin with the premise "I can think of *something* that won't work . . ." become even more foolish when working solutions have already been published. Nonetheless, this sort of nonsense passed for scientific criticism of APM concepts just a decade ago.

The moral of the story: Engineers seek designs that substantiate concepts, scientists seek tests that disprove theories, and it's a mistake to attempt to disprove an engineering concept by proposing and rejecting a design that won't work.

Simple, Specific Theories vs. Complex, Flexible Designs

In developing designs, engineers favor simplicity but accept whatever complexity a task may require. In developing theories, scientists favor simplicity and won't consider a theory if it's too complex.

Physicists see the Standard Model as uncomfortably complex and in need of replacement, yet its equations can fit on a single page. By contrast, the Saturn V launch vehicle, with its six million components, generated hundreds of tons of documentation, yet engineers saw this not as fatal inelegance, but as a challenge for innovative management.

Science likewise has no use for a theory that can be adjusted to fit arbitrary data, because a theory that fits anything forbids nothing, which is to say that it makes no predictions at all. In developing designs, by contrast, engineers prize flexibility—a design that can be adjusted to fit more requirements can solve more problems. The components of the Saturn V vehicle fit together because the design of each component could be adjusted to fit its role.

In science, a theory should be easy to state and within reach of an individual's understanding. In engineering, however, a fully detailed design might fill a truck if printed out on paper.

This is why engineers must sometimes design, analyze, and judge concepts while working with descriptions that take masses of detail for granted. A million parameters may be left unspecified, but these parameters represent adjustable engineering options, not scientific uncertainty; they represent, not a uselessly bloated and flexible theory, but a stage in a process that routinely culminates in a fully specified product.

In any complex project, large-scale funding always precedes a detailed design. Flexibility and design margins enable confidence that designers will be able to fill in the blanks. In Apollo, again, engineers understood what sorts of questions could be deferred, and their colleagues could review and revise their decisions. Options, flexibility, margins of safety, and conservative models—these are parts of an art of design that requires a mode of perception and judgment that forms a natural complement to science, yet requires different skills.

The moral of the story: Beware of judging designs as if they were theories in science. An esthetic that demands uniqueness and simplicity is simply misplaced.

Curiosity-Driven Investigation vs. Goal-Oriented Development

As we saw in the parables of elephantology and automobility, the organizational structures of science and engineering—effective ways to coordinate day-to-day work—aren't interchangeable.

In science, independent exploration by groups with diverse ideas leads to discovery, while in systems engineering, independent work would lead to nothing of use, because building a tightly integrated system requires tight coordination. Small, independent teams can design simple devices, but never a higher-order system like a passenger jet.

So long as research still centers on learning to make new components—in a nascent engineering field, like much of nanotechnology today—designing and building complex systems will remain a task for the future. Later, however, when a rich enough set of components can be made and assembled, the door to systematic engineering stands ready to open.

Crossing this threshold requires not only a different mode of thinking (in the ways outlined above), but also a different way of organizing work—not to replace investigator-led scientific research, but to enrich its value though applications. In inquiry, investigator-led, curiosity-driven research is essential and productive. If the goal is to engineer complex products, however, even the most brilliant independent work will reliably produce no results.

Indeed, without an engineering approach to exploring potential, the very recognition of opportunities becomes unlikely. To architect systems concepts in a new domain and to understand how to move forward toward their realization—a trained scientist may accomplish this science-intensive intellectual task, yet the task itself is, in its essence, an abstract and high-order form of engineering.

The molecular sciences are in this situation today. Advances in atomically precise fabrication have opened the door to systematic AP engineering, yet within the sciences themselves this engineering potential has been barely perceived. The pitfalls outlined in the sections above suggest some of the reasons.

The moral of the story: Beware of approaching engineering as if it were science, because this mistake has opportunity costs that reduce the value of science itself.

APPLYING THE ENGINEERING PERSPECTIVE

Scientific inquiry requires an intellectual discipline that, in its ideal, seeks simple, universal explanations while shunning complex, narrow hypotheses; one that seeks both precision and unpredictability; one that seeks tests that can narrow competing hypotheses toward a single truth, while recognizing that a single, precise, universal truth can never be known with certainty. Scientific inquiry, by definition, seeks to expand human knowledge.

Drawing on established knowledge to expand human capabilities, by contrast, requires an intellectual discipline that, in its fullest, high-level form, differs from science in almost every respect.

System-level design seeks flexible models and accepts complexity, accepts imprecision and buries it in margins of safety, recognizes unpredictability and seeks to avoid it, embraces competing options and seeks to add more, all in a world in which questions have a range of satisfactory answers and testing can establish a narrow yet valuable truth: "This one works."

Engineering and science share knowledge and serve one another, often smoothly and with clarity of purpose, and their essential, defining aspects—design and inquiry—often mesh closely within a single program, laboratory, or mind, regardless of which flag is flying.

When a new engineering discipline emerges from science, however, the very nature of the new challenges and opportunities may escape recognition, because scientists' minds will be focused on problems of a different kind. Opportunity costs grow as the science advances, as they have today in the molecular sciences.

As an engineering field emerges, a new world comes into view. That world can't be seen through the lens of science alone, yet without science, engineering is blind. A clear view of the prospects (when one can be had) requires a kind of binocular vision, a view through the lenses of science and engineering together. This binocular perspective on the landscape of technological potential may sometimes reveal landmarks at a distance, and when the view also reveals a path forward through accessible objectives, it may help research leaders choose their next steps.

In this effort, what might seem most problematic is the longer-range view, the attempt to identify landmark objectives and sketch some of their promise. APM-level technologies are among those landmarks, like a mountain range on the horizon. How much can be seen?

Exploring paths across the timeless landscape of technological potential requires asking what are sometimes unfamiliar science-based questions about technology. A crucial step is to understand what kinds of questions can be asked and answered, and to understand the nature of the methods, answers, limitations, and power of exploratory engineering.

CHAPTER 9

Exploring the Potential of Technology

SCIENCE ASKS, "How can we discover *new knowledge?*" Engineering asks, "How can we deliver *new products?*" Exploratory engineering, however, asks a different, less familiar question: "How can we apply *existing* knowledge to explore the scope of *potential* products that cannot yet be delivered?" In other words, exploratory engineering applies existing scientific and engineering knowledge to explore the landscape of potential technologies.

The history of spaceflight began with exploratory engineering. Advances in physics, chemistry, and engineering principles had provided the essential knowledge, and a man with a vision soon set out on a journey of ideas, a journey guided by quantitative dreams. This first expedition returned with a rough yet accurate map of the landscape of spaceflight, a map that outlined paths to technologies that now have emerged as physical realities along with paths that lead beyond, toward visions that still guide today's plans.

The journey began long ago.

SPACE: 1899

In 1957, the Soviet Union shocked the world by launching Sputnik, an "artificial moon" that arced through evening skies over every backyard nuclear fallout shelter in the United States of America.

Spaceflight research, though, had begun long before. Russian research pioneered the fundamental principles of spaceflight, establishing the equations of rocket dynamics, identifying the highest energy fuels for rocket propulsion, developing the concept of multi-stage rockets, and exploring the essential requirements for sustaining human life in deep space. This seminal work laid out the foundations of astronautics, and more. Even the earliest Russian research looked beyond Earth-orbiting rockets and lunar voyages to envision human settlement of the Solar System.

This early Russian space program began under Tsar Nicholas I more than a century ago. It consisted of the work of one man—a self-taught provincial schoolteacher in Kaluga, working at home in his spare time to explore and give substance to a vision that was planted in his mind by the French science fiction writer, Jules Verne. The vision was space-flight and the man was Konstantin Tsiolkovsky.

In 1896, after a decade or more of reading, thought, and calculation, Tsiolkovsky completed his first, landmark work, wrote it up as a technical paper, and submitted it to the Russian journal *Scientific Review*, where it gathered dust for half a decade. Tsiolkovsky's paper, "The Exploration of Cosmic Space by Means of Reaction Devices," finally reached print in May 1903. Later that year, in December, the Wright brothers flew their first motorized glider to altitudes as high as ten feet.

Tsiolkovsky's achievement presents a puzzle. How could a provincial schoolteacher—working alone, by lamplight, before the first airplane took flight— become known to later generations as the "Father of Astronautics"?

The answer to this historical question still matters today, because Tsiolkovsky's method has enduring power. Call it "exploratory engineering"; as applied by Tsiolkovsky a century ago, this method of study showed that rocket technology could open a world beyond the bounds of the Earth. Applied today, this method (which guided my work at MIT) shows that atomically precise technologies can open a world beyond the bounds of the Industrial Revolution.

Exploratory engineering exploits available physical knowledge and engineering methods to explore the potential of physical technologies. As with all of engineering, the process of exploratory engineering cycles between design and analysis, between expansive imagination and restrictive calculation.

Exploratory engineering thus has much in common with engineering that has more immediate aims. What makes exploratory engineering different is its purpose. Standard engineering delivers artifacts; exploratory engineering delivers knowledge. Standard engineering must respect not only physical law, but also the constraints of manufacturing, the limits of what can be made in a particular historical era. Exploratory engineering, in its purest form, respects only the eternal constraints of physical law itself.

The extent of available knowledge determines how far exploratory engineering can reach. In 1896, when Tsiolkovsky presented his analysis of reaction devices (that is, rockets), physical knowledge was a fraction of what we have today. Henri Becquerel discovered radioactivity that year and Wilhelm Röntgen had discovered X-rays the year before. Electromagnetic waves were known, but Guglielmo Marconi's first radio station had yet to be built. The nature of matter itself was a mystery. Because physicists had no conception of quantum mechanics, they knew next to nothing about the fundamental nature of materials and molecules— indeed many doubted the very existence of atoms. (Chemists knew better, aside from a few die-hard skeptics.)

Nonetheless, Tsiolkovsky had knowledge enough for what he did. He applied Newton's laws of mechanics to describe the principle of reaction devices ("for every action . . ."); he applied Newton's law of gravity to quantify the challenges of reaching orbit; and he relied on experimental data to identify high-performance propellants, the combinations that burn to release the highest energy per unit mass. Tsiolkovsky suggested that liquid hydrogen and liquid oxygen would be the best propellants, and space systems engineers use them today.

Tsiolkovsky had no need for a lathe or a drill because he produced his results using the traditional tools of design engineering: pen and ink. He didn't develop rockets; instead, he developed the rocket equation. He didn't build liquid-fuel rocket engines; instead, he showed that liquid fuels would offer the highest performance.

He offered no detailed designs—those came later—but he showed how the basic problems could be solved, and thus charted a path deep into the world beyond the sky: "Man will not always stay on Earth; the pursuit of light and space will lead him to penetrate the bounds of the atmosphere, timidly at first, but in the end to conquer the whole of solar space."

From his home in Kaluga, building on known physics and chemistry, Tsiolkovsky constructed a vision of technological potential that continues to unfold into physical reality.

A proper biography of Tsiolkovsky would mention that he lost his hearing at age ten, which ended his formal education and set him on a course of self-education; that he married and raised a family; that he expounded a truly cosmic philosophy of human destiny, was ignored for many years, and supported the Bolsheviks, who first jailed him in Lubyanka Prison, then celebrated his vision, electing him to what later became the Soviet Academy of Sciences and awarding him a lifetime pension.

A proper history of the origins of spaceflight would note Tsiolkovsky's limited influence outside Russia and the later, independent roles of Robert Goddard in the United States and Hermann Oberth in Germany with their concrete technical contributions: building actual, physical, liquid-fuel rockets. The history then would tell of the rise of rocketry on the tides of war, led by visionaries sharing Tsiolkovsky's dream of the conquest of space. The leading example was Werner von Braun, the German scientist who led the German V2 rocket project that rained bombs on London from the edge of space, envisioned a detailed plan for interplanetary exploration (*Das Marsprojekt,* published as *The Mars Project* in 1953), was ordered to refrain from launching a satellite for the United States in 1956, and at last led the team that developed the Saturn V boosters that launched men to the Moon.

My attention here, however, centers not on the people themselves nor or the tangled history they lived, but instead on the methods of thought they developed and what their example teaches about the discipline of probing potential technologies.

LEARNING FROM SPACE SYSTEMS ENGINEERING

Exploratory engineering established the potential of rocket-based spaceflight technology decades before anyone could build the rocket hardware required. It was systems-level engineering, however, that translated this potential into detailed reality.

The history of space systems development shows how practical engineering can emerge from concepts beyond current practice, giving substance to ideas of architectural scope and abstraction, ideas born of imagination, then tested against the principles of engineering and physical law. Starting with the exploratory, conceptual birth of space flight, history shows how design, stepwise refinement, redesign, and further refinement can set projects in motion that in the end produce working machines.

From liquid-fuel rockets to Sputnik to rovers on Mars, the history of space technology illustrates the flow of systems engineering. Behind the practice of space systems engineering today stands a body of methodologies, standards, traditions, and institutions, animated by a culture that took form in the mid-twentieth century. I joined this culture while at MIT and learned from its practitioners the art of architecting audacious, quantitative visions, always discarding all but a few.

Exploratory engineering has much in common with the initial stages of systems engineering—it applies similar methods of high-level design and analysis, but aimed at a somewhat different objective. Both endeavors begin with high-level abstractions that define patterns of functionality and then ask what configurations of subsystems could be used to achieve an overall functional goal; in space systems, one might seek a configuration of structures, fuel tanks, and engines that could be used to boost a ten-ton satellite into geosynchronous transfer orbit. However, rather than considering launch vehicles built using available aluminum alloys, exploratory engineering might examine the performance of vehicles built

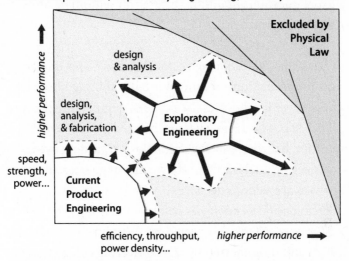

FIGURE 2: Exploring the Technological Implications of Physical Law

using atomically tailored carbon materials that provide fifty times greater strength per unit mass (the resulting performance improvements are huge).

ENGINEERING WHEN THE PRODUCT IS KNOWLEDGE

Standard systems engineering seeks designs that can lead to a reliable, competitive product; exploratory engineering seeks designs that can lead to a reliable, significant conclusion. They ask and answer different kinds of questions.

Understanding the answers to exploratory engineering questions requires understanding the nature of the questions themselves, both what they *are* and what they are *not*. In particular, because exploratory engineering questions address the timeless landscape of technological potential, their answers don't depend on unknown, contingent facts. The answers to exploratory engineering questions can have implications for potential futures, just as the newly discovered physics of nuclear fission had implications for potential outcomes to World War II, yet having implications *for* the future doesn't make them questions *about* the future. Events, whether past or future, do not change physics.

The usual questions *about* the future can seldom be answered with confidence. Here are a few:

- *What path will technology actually follow?* This question asks for predictions of human events, inviting informed guesses, yet it is beyond the scope of engineering analysis.
- *Which specific technologies will succeed in the market?* This question likewise asks for predictions regarding events; here, engineering analysis can help to inform guesses, but the answers are seldom decisive.
- *Which specific technologies will ultimately offer the highest performance?* This may seem like an engineering question, yet it can't be answered by an engineering calculation; an answer would require impossible knowledge of every possible alternative that might someday be discovered.

Put "technology" and "future" together, and the mind is apt to muddle together knowledge based on physical principles with speculative talk that seems superficially similar—unanswerable questions, dubious answers, and with these the kind of hype that evolves in marketing, media, and popular culture.

Exploratory engineering, however, asks a special, carefully chosen kind of question—a kind that doesn't depend on future unknowns, and isn't made difficult by competitive pressures. By design, these questions can be answered using information that's already available. Exploratory engineering then makes answers readily accessible by designing systems in an extraordinarily conservative way that can relieve pressures on both design and analysis. The questions ask only for lower bounds on potential performance, and hence can build on layers of worst-case analysis, rather than asking what would be best. As a consequence, the answers describe systems so conservatively designed that they are virtually guaranteed *not* to describe a future competitive product.

What is the use of such limited knowledge? Think of competitive engineering as adding detail to a chart showing coastlines and rocks, sea lanes and harbors, cities and roads—detail that adds up to a precise, textured description.

Think of exploratory engineering as sketching the coast of a new continent, tracing its outlines and marking rivers that lead to fertile valleys.

A rough map can guide further exploration by charting regions worth reaching. In engineering, a rough map and direction—*Up into space, with multi-stage rockets!*—can set the human race on a journey that had been beyond dreams. Decades of travel may be required to reach a new land, but the end is sometimes in sight from the start.

In a journey of ideas and new technologies, asking for details of the next decade's machines is usually fruitless—useful to consider as an aid to planning, but not as a prediction. Yet a different, exploratory question, also useful for planning, may ask for no more than an outline of what could be done. In spaceflight, this question was asked and answered, again and again, and the earliest questions led to the work that answered the rest.

Engineering with an Exploratory Twist

The Three Cardinal Rules of Exploratory Engineering

Exploratory engineering is the art of applying scientific knowledge and engineering methods to explore the potential of future technologies. Three rules describe its essential methods:

- Rule 1: Explore systems of kinds that current tools can't build.
- Rule 2: Ask only questions that current science can answer.
- Rule 3: Think like an engineer.

In other words, explore areas not already mapped by developers, stay on solid ground, and apply both imagination and analysis to explore potential objectives, problems, and solutions.

> Finding a permissible stress level which will provide
> satisfactory service is not difficult. Competition forces a
> search for the highest stress level which still permits
> satisfactory service. This is more difficult.
>
> —*Standard Handbook of Machine Design,* third edition

Conventional engineering aims to provide competitive products, exploratory engineering aims to provide confident knowledge, and these radically different objectives call for different methods.

Both kinds of engineering begin with a purpose, an intended function that defines requirements, and both begin with system-level conceptions that then define subsystems with their own requirements (and so on, until the recursion hits bottom). Both methods advance toward their goal through cycles of design, analysis, and refinement, but their different goals call for different design criteria, a different approach to analysis, and a different degree of design refinement. The pressures of competition play a key role.

Competition leads engineers to seek optimal choices (in the *Standard Handbook of Machine Design,* for example, "a search for the highest stress level"), while in exploration, a more typical problem would be to find "a permissible stress level" (which can be, as the *Handbook* remarks, "not difficult"). In exploratory engineering, however, optimization is optional, and simplicity and conservative design trump performance.

- -

Engineering to Build Knowledge

Contrasts between production-oriented and exploratory engineering:

Competitive products	Confident knowledge
Production details	System analysis
Manufacturing limits	Modeling limits
Efficient design	Conservative design
Cost of production	Cost of analysis

- -

Exploring what could be easily reached with prospective technologies isn't the same as trying to squeeze more from technologies already in hand. Discussed in this book, is the low-hanging fruit of APM-level technology.

Engineering Reliable Products vs. Engineering Reliable Knowledge

In production-oriented engineering, success means delivering a reliable physical product, a physical system that consistently performs as it should. In exploratory engineering, however, success means delivering a body of knowledge that performs as it should. These different products call for different processes.

In developing reliable products, engineers provide ample margins of safety to counter unexpected events, defects in manufacturing, and inaccuracies in the design and analysis itself. To keep passenger aircraft in the air, engineers also add backup systems to counter unexpected failures. Finally, at the end of the design and production pipeline, engineers gain yet more confidence by stringent testing.

In exploratory engineering studies, by contrast, where the design itself is the product, "reliability" means confidence that a system-level design could be refined into any of a range of specific implementations that would fulfill what the analysis has promised. To gain this confidence, an exploratory engineer applies large safety margins to accommodate uncertainties and to simplify design and analysis.

In the usual sort of production-oriented development, competitive pressures push engineers toward risky designs and narrow margins, and striking the right balance presents challenges that pervade the design process. Without these pressures, exploratory engineering studies can indulge in the luxuries of design simplification and can add layers of padding to absorb the impact of surprises.

Cost and Performance of Products vs.
Cost and Performance of Design and Analysis

Relaxing performance (compared to what might be optimal) helps explorers find realms of extraordinary performance (compared to anything available now). On the landscape of potential technologies, location matters more than refinement, because new locations are realms with new rules.

For example, in 1960, after half a century's progress, an efficient aircraft could stay aloft for no more than tens of hours without refueling, yet even the earliest satellites could stay aloft for millennia. Satellites had quite literally entered a realm with new rules. Likewise with the future potential of manufacturing. Simple, conservative implementations of advanced APM can far outperform the most refined modern means of production, again because the new realm has new rules.

Because location matters more than refinement, the cost structure changes.

In product-oriented engineering, the costs of manufacture and operation typically dwarf the cost of design, and as a consequence, large investments in design can bring great rewards. In a production run of one hundred aircraft, for example, a bit of design work that shaves a kilogram from airframe mass can save a million dollars in fuel costs. Likewise, competition speaks against padding designs with needless margins of safety—excessively strong and heavy wings, for example, or excessively powerful, heavy, and fuel-hungry engines.

In exploratory engineering, by contrast, design cost is the only cost, and because the endpoint of an exploratory design is a systems-level analysis, even the cost of design is low. Without production, the cost of materials for manufacturing is zero. Without products in use, the cost of operation is zero. The cost of adding a broad margin of safety then falls below zero, because including broad margins is a cheap way of building confidence into a design analysis.

In the case of APM systems, for example, discarding a factor of one hundred in potential productivity makes little difference to any conclusion that matters here, and the freedom that comes with this enormous design margin works wonders for confidence that constraints can be met.

In exploratory engineering, the cost of system-level design and analysis is the whole cost of making the product—but what is the cost of using it? Because the product is knowledge, the primary cost of using it is someone's time and attention, the cost of understanding and judging its content. Like the cost of design itself, this cost also falls when designs are presented at a systems level and padded with large margins of safety. Thus, exploratory engineering methods economize the intellectual costs of knowledge from end to end.

CHARTING PATHS FROM HERE TO APM

The timeless landscape of potential technologies embraces both products and the means to make them. Looking back, we can see that today's technologies were enabled by yesterday's technologies in an unbroken chain of progress that leads back through the Industrial Revolution, then back to hammers and iron, and onward back to beginnings with tools of stone and stone-carved wood. At every stage, fabrication technologies enabled their better successors. Each fabrication technology marked a place in the landscape where performance was best measured, not by strength, or power, or computer speed, but by the scope of what could be made, by the entire content of new territories brought into reach.

And within those new territories, what mattered most in the long run were new technologies that opened further territory, which is to say, the new technologies that themselves were useful for making new things. AP fabrication technologies will increasingly have this character: Along paths toward APM-level technologies, improvements in AP fabrication technologies will enable improvements in those same technologies.

How can exploratory engineering help identify paths? First, by treating a relatively narrow set of fabrication capabilities as implementation constraints (on accessible materials, components, and assembly methods), then by treating expanded fabrication capabilities as design objectives, seeking to expand the range of accessible materials, components, and assembly methods. Each set of potential fabrication capabilities marks a place on a path, and a design for a system built with this set, and adding to it marks a potential step forward. The aim, then, is to explore lines of development in which each level of fabrication technology enables the next, defining a series of closely spaced stepping stones.

Appendix II surveys a smooth terrain for paths that lead from today's level of laboratory capabilities to APM-level technologies through incremental steps. The length and pace of actual steps will depend on the quality of design insights and coordinated effort.

Terrain slopes upward from current laboratory capabilities, which include a surprisingly fast-growing range of atomically precise fabrication technologies with roots more than a century deep. The current state of the art embraces structures on a scale of millions of atoms—far larger than the programmable machines in the cell—together with materials with performance in the range of high-grade engineering polymers and even AP components stronger and stiffer than steel.

The limit to what can be done with these materials isn't fabrication, per se, but the challenges of modeling, component design, and refining suitable systems-level engineering concepts. Here, the divergent perspectives of science and engineering have left a gap open, not blocking progress, but slowing its pace.

Today's techniques stand at one end of an upward gradient of technologies that leads to APM-level fabrication capabilities. Natural paths forward augment conventional self-assembly by adding soft mechanical constraints, shifting (later, and further up the gradient) toward greater reliance on positional control and less reliance on complex, matching surfaces. The rewards include smaller, simpler, more densely bonded building blocks that combine to form stronger, stiffer, finer-grained materials; together, these improvements enable increasingly straightforward design.

Each step up the slope can yield a wider range of higher-performance products (materials, therapeutic agents, digital devices, and so on) in quantities ranging from billions of nanoscale units (a microgram, perhaps) to tons of material. Along these paths, however, advanced, scalable APM-level systems are late developments; they will require large, well-orchestrated arrays of devices that produce nanoscale components and pass them to larger machines for assembly. Appendix II surveys paths in more detail.

The next chapter looks toward a region high up the slope—toward advanced APM-level technologies and systems, the basis for developments that merit the name "radical abundance."

PART 4

THE TECHNOLOGY
OF RADICAL ABUNDANCE

CHAPTER 10

The Machinery of Radical Abundance

THE PREVIOUS CHAPTERS have provided conceptual fragments, and it's now time to put the pieces together and see what an atomically precise manufacturing system might look like, from end to end, and then turn to the question of how well APM is now understood. (Technically inclined readers will find useful notes on the molecular-level physical principles of APM in Appendix I.)

Earlier, I compared potential APM systems to printers, packaged as tabletop appliances for making small-scale products. Here I will describe how small mechanisms can be organized to make larger products, the size of an automobile, for example. At this scale, we can picture a system configured as a demo factory, complete with windows for watching how the machinery works.

LOOKING BACK FROM THE PRODUCT END

Picture yourself standing outside the final assembly chamber of a large-product APM system and looking in through a window to view the machines at work in a space the size of a one-car garage. (There's no magnification here.)

To the right, you see an exit door for products ready for delivery. To the left, you see what look like wall-to-wall, floor-to-ceiling shelves, with each shelf partitioned to make a row of box-shaped chambers. In the middle of the garage-sized chamber in front of you is a movable lift surrounded by a set of machines.

The machines look uncommonly sleek, yet very familiar. They resemble machines in an automated factory, with robotic arms programmed to swing around, pick up components, and swing back to snap the components together. The machines look like this because they are, in fact, machines in an automated factory and because machines that perform similar motions often have similar shapes and similar moving parts. Because they are made of materials better than steel, however, they can be faster, lighter, and more efficient.

Looking back at the wall on the left, you can get a clear view into several chambers that happen to be at eye level and near the window. Each smaller chamber contains machines with swinging arms, and the overall setup inside looks like a scale model of the larger chamber, complete with a rear wall with wall-to-wall, top-to-bottom rows of yet smaller chambers. It's hard to see in detail what these small chambers-within-chambers contain, but they seem to hold a tiny yet familiar set of machines mounted in front of a rear wall with rows of yet smaller chambers.

With the press of a button, the machinery kicks into gear.

At first nothing seems to happen, but in less than a minute the large machines in front of you start to pick up parts as they pop out of the chambers in the wall at the left, moving these parts to the platform in the center where the first parts are clamped, and the rest snap together. As the machines put the parts together, a familiar product takes shape, an automobile, different in almost every detail from those built today, yet having a form that reveals the same function.

Each part takes several seconds to put into place and new parts slide out of the chambers at a corresponding rate, each chamber delivering a component every few seconds. To the left, inside the closest chamber, you can see the machines working inside. These miniature machines seem to be performing similar tasks, but at a rate of several cycles per second, their motions are almost too quick to follow. It's easy to guess what's happening in the yet-smaller chambers farther back, yet the motions there are no more than a blur.

In the main chamber the work is complete in less than a minute. The door to the right then unseals and opens, and a car moves out into a receiving area, sealed in what looks like a plastic sleeve. A moment after the door reseals, the sleeve is pulled back for recycling and the process is done. (This exit maneuver is part of a cycle that prevents contaminants from entering when the product exits.)

If you wanted to buy this minutes-old car as a souvenir, it wouldn't cost more than a good restaurant dinner (and with a discount for watching the demo). Unfortunately, owning yet another light-weight, high-performance, zero-emissions vehicle would just add to clutter at home. Instead, at the touch of another button, the car rolls into a neighboring machine where its parts are recycled.

LOOKING DEEPER INTO THE PROCESS

The last steps of the APM process are so ordinary that, with just a few adjustments, today's factory robots could do the job, putting macroscopic parts together as they do in manufacturing plants every day. At this end of the process, the unique aspects are the quality, structure, and performance of the components and the way they're crafted to fit together smoothly, tightly, and quickly, without the need for welds, bolts, epoxy, or rivets. Looking deeper into the process, however, shows something different.

The smaller chambers with their smaller, higher-frequency machines tell part of the story. As Chapter 5 showed, mechanical scaling laws apply to space and time together. In machines with parts of similar shape that move at equal linear speeds, smaller parts will travel shorter distances in proportionally shorter times, and what's more, it turns out that all of their dynamical properties (stress, strain, vibrational frequencies, and so on) scale in the same proportion. Thus, smaller machines can perform similar motions, assembling smaller parts at higher rates. This is why in the demo factory the view back into progressively smaller chambers leads to machines moving in a blur of motion, at frequencies of tens, then hundreds, then thousands of cycles per second as the sizes decrease.

In this conceptual demo factory, how far does the series of chambers reach in terms of actual length? If each step back shrinks chamber sizes by a factor of ½ (for example), then successive layers will have thicknesses of 1, ½, ¼, ⅛, and so on, in a series that adds up to 2 (or to pick another ratio: $1 + \frac{1}{3} + \frac{1}{9} + \frac{1}{27} + \ldots = 1\frac{1}{2}$). In other words, the entire physical length of the sequence, no matter how long when counted in terms of layers, is about the same as the length of the outermost chamber.

How much time does the assembly process require, from end to end? The outermost stage must allow time for each arm to perform several motions that can all be completed in, perhaps, thirty seconds. Each series of motions removes an assembled component from a smaller chamber, so in order to keep pace, each smaller chamber in the previous stage must complete several assemblies in this same thirty-second interval— and with smaller, higher-frequency machines, it can. Scaling laws naturally lead to a balanced result in which each layer of chambers handles equal amounts of mass in equal times, using mechanisms that all move at equal linear speeds, regardless of size.

The net result is simple. In this kind of convergent assembly process, components flow from end to end at a constant average speed. If the components spend, say, thirty seconds during assembly in the final chamber, and if all the layers of the smaller chambers, taken together, double the length of the system, then the total delay will be roughly a minute—that is, if the process starts with blocks substantially larger than molecules.

In this conceptual-demo APM system, tracing the paths of the parts back toward their sources would require a jeweler's loupe, then a microscope. With the aid of progressively greater magnification you'd see a sequence of smaller and smaller machines and chambers leading ultimately back to machines that build components by assembling atomically precise microscale building blocks.

These smallest machines are much like the ones we saw in the outermost chamber, similar in terms of shapes, motions, and tasks despite almost a million-fold reduction in size, though different in many details (the implementations of motors and bearings, for example).

Deep in the nanoscale-size range, however, engineering encounters ultimate, atomic constraints on the size of even the simplest devices. Gears, shafts, and bearings, for example, can't be made smaller than a few nanometers in diameter, simply because mechanical components must contain enough layers of atoms to provide suitable shapes, surfaces, and mechanical properties for the devices to function. And note that only mechanical devices scale in the way described here; electronic devices, by contrast, exhibit far-from-classical behavior well before reaching this size range.

Tracing back further, to the source of the microscale blocks, would reveal machines that look nothing like programmable robot arms. This deep in the nanoscale world, independently programmable machines become impractical simply because the smallest possible computers become—by comparison—impractically large. Because digital systems are already built with nanoscale components, they can't be scaled as far down as machines. Today, the smallest practical factory machines are larger than a computer chip; in the nanoscale world, by contrast, the sizes are reversed because machines can scale down by an enormous factor while computers remain relatively large, energy-hungry, and slow. Thus, in a practical APM factory architecture, the disproportionate size of computers calls for using machinery of a different kind: simpler, computer-free, and more efficient.

CROSSING THE MICROBLOCK THRESHOLD

The natural threshold scale for this change in the style of machinery (and a natural place for a gap in the APM supply chain) is in the range where machines build larger-scale components by assembling microscale blocks. Blocks in the microscale size range can be large compared to ordinary molecules—hundreds to thousands of times larger in diameter and millions to billions times larger in volume—yet at the same time can be smaller than the wavelength of visible light, small enough that a clear view of one would require an electron microscope.

Like droplets in an inkjet printer, microblocks can be combined to form endless, intricate patterns from small bits of material. A better analogy than ink droplets, however, would be blocks of the sort found in high-end Lego sets, which include not only blocks in different shapes and colors, but also blocks that provide intricate, functional parts such as motors, gear trains, sensors, and computers.

Microblocks would have wider applications than the Lego analogy suggests. Microblocks made of super-strong materials can be designed to slide into alignment with atomic precision and then bond tightly on contact to form super-strong parts; microscale blocks can contain digital circuits as complex as a computer processor core, and devices like these can be assembled to make supercomputers far beyond today's state of the art. Within the microblock-size range, larger blocks enable faster assembly, while smaller blocks enable finer-grained customization, and there's no reason to limit choices to a single size. There are endless options for organizing parts and production methods, and the most practical choices will vary with the purpose.

Because the microblock threshold is a natural place for a gap in the supply chain, exploring our demo APM factory further back would lead, not to more production machinery, but to what amounts to a warehouse, a stockpile of microblocks delivered from another facility.*

It's time to consider systems that can produce a wide range of microblocks, where machines all the way down at the nanoscale produce atomically precise microscale components. Here it makes sense to start with molecules, and then work forward and upward in scale.

FROM MOLECULES TO MICROBLOCKS

How can an APM supply chain produce AP microblocks from raw materials? The process involves a series of stages, from raw materials to

---

* Reasons for separating the upstream processes include the logistics of raw materials supply, refining, and solid waste management, together with the technical requirements of high-throughput, molecular-level operations (such as managing heat dissipation and chemical-intermediate recycling), and the advantages of efficient, specialized hardware for all of the above.

pure, refined materials, then from bound molecules to monomers, and finally, to microblocks.

Consider the process from raw materials to purified feedstocks. To make exploratory engineering tractable, it's best, when possible, to push complexity outside. The process of obtaining and refining raw materials entails intimate and inherently messy contact with the stuff of nature. We already have ways to do that by means of familiar industrial processes, so for the moment let's push the complexities of things like rock, petroleum, and seawater outside the atomically precise box and consider an APM supply chain that relies on current technologies to provide common, commercial-grade chemical substances. But note that moving novel—and cleaner—technologies further upstream is a natural step in upgrading an APM supply chain.

These commercial-grade inputs must provide all the elements that will appear in the outputs, and preferably little else. I've mentioned carbon and hydrogen in connection with output materials, and light hydrocarbons (whether from petroleum or a renewable source) can provide these elements. Air can provide oxygen and nitrogen, while both silicon and aluminum can be delivered as inexpensive water-soluble compounds.

The next step, purification, takes a step toward atomic precision by excluding contaminants from a fluid feedstock stream. A multi-stage purification system (itself built with APM-level technology) can reduce contaminants down to a level approaching zero. Systems of this kind are widely used in chemical engineering and drive impurity levels down by a constant factor per stage, which is to say, exponentially. The resulting feedstock materials consist of simple molecular structures, and any specific molecular structure is, by definition, an atomically precise object. In APM this is where atomic precision begins.

To maintain this initial atomic precision from purified feedstocks to activated monomers, the next step in the process is to bind the incoming molecules to sites on larger molecular structures that also serve as mechanical parts. Picture a conveyor belt, a chain of containers each with a site that binds a feedstock molecule from a purified fluid, and in a particular orientation.

Beyond this point, the feedstock stream itself is atomically precise, consisting of a chain of precisely built containers, each holding a specific, oriented, molecular structure. The messy disorder of the familiar world of chemistry and biology has been left behind—outside the walls of the box—with all the advantages that this implies for simplifying system design, analysis, and function. Further operations can maintain atomic precision, provided that each step has a definite, reliable outcome.

The downstream mechanisms guide molecules through a series of encounters with other molecules (some best described as molecular tools), thereby directing chemical steps that prepare reactive bits of molecular structure—bound, activated monomers—that can be joined together to make covalently bonded objects. Each step maintains the initial atomic precision of the feedstock molecules themselves.

The products can be extremely diverse. Machines that guide the bonding of the smallest monomers—just an atom or two—can construct patterns of atoms like those formed by means of less-controlled processes, and, of course, many more. In other words, they can make the full range of materials used in technology today, but in addition can craft them with intricate shapes and internal structures to produce an endless range of high-performance AP components.

How productive can molecule-processing machines be? Because of mechanical scaling laws, mechanisms can easily guide streams of molecular encounters at rates of ten million per second, and design experience suggests that mechanisms in the ten-million atom range can include all the devices needed to do the job. These rough numbers suggest that a typical molecule-processing machine of the sort just described could deliver its own mass in the form of products in about one second.

Note that a machine that instead took minutes to produce this much product would still be enormously productive, hence this estimate is far from critical. If a system were to use even one hundred times as many machines as seems necessary, the system would still be radically productive. Thus, adding layers of conservative design would have little effect.

As we move downstream, from monomers to functional microblocks, larger atomically precise structures (now with many atoms) can be joined together like monomers, but forming many bonds at once. And when yet larger blocks come together (a few nanometers wide is enough), universal molecular attractive forces become strong enough to form stable connections between matching surfaces with no bonding required at all.

Fifty steps of monomer buildup and convergent assembly are enough to span the gap between small molecular fragments and AP structures built on a micron scale—microblocks.

A block half a micron wide can hold ten billion atoms, enough to comprise ten million components of one thousand atoms apiece. Blocks with this scale and potential complexity are large enough to provide components that serve any of a range of functions, including structural components of course, but also devices like electric motors, computer processors and memories, and sensors and actuators of many kinds, all of which can be packaged as plug-together modules.

Specialized microblock production lines can also produce modular components for systems that perform molecular purification, binding, and monomer activation along with other components that can be plugged together to make systems that move and join monomers, and then larger nanoscale chunks. A further kit of components could be used to build larger machines across a range of scales, including components that can be assembled to build structures, motors, interfaces, and controllers for larger-scale factory equipment, from microns to meters in size.

In other words, a sufficiently diverse set of microblock production lines could produce all the components required to build similar microblock production lines along with equipment for factories like the one already described, factories that can build things like automobiles and factory equipment.

A NOTE ON ENERGY REQUIREMENTS

The molecular end of an APM process requires input energies on a chemical scale simply because the operations involve chemical transformations of molecules. The energy required for purification is low by comparison, as is the energy cost of driving the mechanical motions. The greatest energy costs arise in monomer preparation and bonding, along with the early stages of processing to build larger blocks.

What sources contribute to the energy cost? First, between input and output there is often an inescapable requirement that results directly from thermodynamics. If aluminum enters in an oxidized form, but exits as a metal, there's an irreducible energy price to be paid, just as there is in conventional aluminum smelting. In addition, there are energy costs in processing bound reactive monomers because each step typically must expend substantial chemical energy to make the change irreversible.

The energy required for steps like these is comparable to the energy released in burning a fuel; for a kilogram of carbon, this amounts to roughly thirty million joules (at electric power prices today, this amount of energy, roughly eight kilowatt-hours, would cost about a dollar). The total, end-to-end energy consumption would vary, depending on the composition of the feedstocks and products and on the efficiencies of the steps along the way. Note that among the important engineering parameters of APM processing, energy consumption is perhaps the hardest to estimate closely. In the next chapter's discussion of economic and environmental costs, I will use conservative estimates; actual energy costs will likely be lower.

Even with good process efficiency, the amount of waste heat released in the early, molecular stages of processing will give reason to place these facilities away from residential areas. Logistics for providing raw materials and removing or recycling wastes point in the same direction.

Early-stage APM processing must pay the energy costs of performing millions or billions of molecular operations per microblock. Once this molecular-level toll has been paid, the road from standardized microblocks to customized, on-demand products becomes easier. For making products of modest size, machines can resemble desktop printers—compact, convenient, ready on demand, and easy to use—and printers are often found where the products are needed, close at hand in the home or a workplace.

GOOD ENOUGH ANSWERS, FOR NOW

How much do we really know today about the potential of APM, the physical, technological potential defined, in the end, by physical law?

At a detailed, atomistic level, the answers today include descriptions of a range of generic, nanoscale components suitable for implementing molecular-level APM functions. Mechanical components have been studied using standard computational methods in molecular mechanics and dynamics, and key molecular transformations have been investigated using density functional methods in quantum chemistry, applied in conservative ways.

Regarding APM systems and their performance, the answers at hand describe architectures based on exploratory engineering and systems-level design and analysis, building on the engineering parameters of novel components as just described. The results are typical of exploratory engineering, providing quantitative descriptions of systems based on conservative assumptions, yielding conservative, lower-bound estimates of potential system performance, where for APM systems performance is defined by productivity, efficiency, and the range of accessible products. These answers describe a set of technologies adequate (and even more than adequate) to support what I've described as radical abundance.

Within the general range of APM-level technologies, the specific level of APM system performance turns out to be remarkably unimportant. The prospects for radical abundance depend on relatively high throughput per unit mass, but whether APM throughput in these terms can exceed current factory throughput by a factor of one hundred thousand or merely one thousand makes little practical difference from today's point of view. Costs of production matter more, yet as shown in the next chapter, even conservative estimates place these costs in a radically low range.

Likewise, the general nature of APM systems and their prospective global impact doesn't depend on any particular process or material; there are many ways of producing each of many high-performance materials, and almost every important product capability, as seen from a practical, human perspective, can be provided in many accessible ways.

Carbon-based supermaterials, in particular, offer extraordinarily attractive performance and are clearly accessible. Conveniently, some of the most reliable methods of modeling molecular structures and dynamics have been developed to study organic, carbon-based materials, and experimentalists have amassed an immense amount of empirical data as well. Because ease of modeling and high performance go together, studies of advanced nanomechanical systems have focused on structures that consist largely of carbon, hence some of the best materials for APM systems and products are also among the best understood. This alignment of prospective performance with current knowledge opens a window on some of the most attractive regions in the landscape of potential technologies.

As a consequence, in an APM context, an exploratory engineer can often make choices that are highly conservative, yet result in exceedingly high performance when compared to the sorts of systems that can be built today. From this advantage, many consequences follow.

———

FROM MOLECULAR PROCESSING to microblocks to product assembly and products, APM-level technologies can only be built using APM-level tools that don't yet exist. Roads from here to there can be seen in outline today, but the roads are neither short nor direct.

In later chapters we will step back to survey the arc of progress to date in atomically precise fabrication and prospects for further advances toward APM-level technologies. Appendix II, in particular, examines accessible paths forward from today's capabilities. The next chapter, however, will continue to explore advanced APM-based production, examining its implications for the costs and capabilities of a range of products. The result will provide a better picture of the nature and potential human impacts of radical abundance.

CHAPTER 11

The Products of
Radical Abundance

APM-BASED PRODUCTION is only as valuable to us as its products and, together with these products, the performance, cost savings, and environmental benefits it can deliver. The same can be said of any form of production, whether it yields stone tools, steel machines, or silicon chips. As we've seen, however, the characteristics of APM-based production point to an impact greater than any of these earlier technologies, with parallels to the digital information revolution, but transposed into the material world. These parallels suggests prospects for a technology-driven transformation that could be faster and more profound than any in history, pushed forward by competitive advantages in cost, performance, and potential speed of deployment.

In the last chapter we explored an APM process that began with commercial-grade raw materials, then produced and combined monomers to build a wide range of microscale building blocks, and finally, elsewhere, in compact factories or desktop boxes, assembled these microblocks to build products up to meters in scale. At its initial and most fundamental level the process relied on guiding the motion of molecules to combine small, atomically precise structures to make progressively larger, equally precise components. This precise control of molecular reactions is what enables APM systems to build structures from the bottom up with atomic precision.

Now we are ready to ask what these capabilities imply and what they will enable. We need to start at the bottom, exploring the performance of extraordinary materials in extraordinary forms, then move up to higher levels: components, products, applications, and costs (in the broadest, yet physical sense of the word). The results of this exploration will offer a concrete view of the potential products of radical abundance, providing a conceptual framework for considering its implications. The resulting picture combines radically low-cost production—in terms of labor, capital, materials, energy, and environmental impact—with products that themselves can be radically better in performance, efficiency, and cost of use.

ASKING THREE FUNDAMENTAL QUESTIONS

"What can be made?"

"What can it do?"

"How much will it cost to produce?"

These are the fundamental questions that any vision of future technologies must answer.

APM's answers hinge upon materials—the nanoscale patterns of atoms and bonds that form the components of APM-level products. As we saw in the previous chapter, the range of accessible materials will be extraordinarily broad. They include the full range of stiff, stable materials made in laboratories today, like carbon nanotubes, silicon carbide, diamond, and oxides and nitrides of silicon and aluminum. With AP fabrication, structures built with similar patterns of bonds can serve as components for products that range from nanomachines to electronic and photonic devices, and beyond.

Researchers can learn much by studying just a few new devices in the laboratory, and without any concern for their cost, yield, or scale of production. In the context of current production technologies, this kind of research provides insights into technologies that may or may not become practical, because on the road to practical applications, making production practical is always part of the problem.

Digital devices offer a good example. Devices made in laboratories shrank to nanoscale size long before any devices that small reached a computer chip; as always, laboratory demos of components—first one, then a few linked together to form a simple circuit—preceded even prototype products. The earliest devices were typically made one by one, sometimes using patterns drawn by scanning a tightly focused electron beam. Industrial production, in contrast, requires devices made billions at a time using patterns projected by light. Even if made with impractical production methods, however, experimental devices can demonstrate the physical principles of devices beyond the then-practical state of the art.

Devices made in laboratories today—highly efficient photovoltaic cells, nanoscale motors and bearings, smaller transistors—likewise demonstrate performance beyond the now-practical state of the art, and again, a key constraint is production.

With high throughput and atomic precision, APM will change the game. With this more powerful manufacturing capability, devices that can be made in the laboratory at a high cost and with low yield can be reproduced at low cost, reliably, on a large scale, and integrated into functional products. Carbon nanotubes, for example, demonstrate higher strength than any other known materials and some kinds are more conductive than copper. Nanotubes of this sort would be widely used if they could be easily made and integrated into materials. This is something APM-based production can accomplish. The situation is similar for nanomechanical devices and advanced photovoltaics, among other examples.

Note that dispersed nanotubes and nanoparticles can pose toxicological hazards (depending on their structures and compositions), but a macroscale product need not shed nanoscale particles, whether or not it's built with atomic precision and nanoscale features. The same is true of APM systems, which could be designed to meet strict toxicological standards at minimal cost.

Thus, the best materials and devices seen in laboratories today establish lower-bound benchmarks for the performance of future APM products.

The results of laboratory demonstrations provide empirical physical facts that can be slotted into exploratory engineering analysis of potential applications.

THE ORIGINS OF RADICAL PERFORMANCE

Many of the products of APM will offer radically better performance than those seen today. There are several reasons, including improved materials (hence better components) and performance advantages that result directly from scaling laws applied to small devices.

In a wide range of products, performance depends on the strength and density of structural materials, and building with stronger, lighter materials can reduce product mass because structural materials make up almost all of the mass of almost every industrial product. As we'll see, production costs will be closely linked to the cost of raw materials and the mass of materials processed, hence reducing mass by building with stronger, lighter materials can yield reductions in cost.

In brief, strong covalent bonding between light atoms can result in materials that are both strong and lightweight (conveniently, some of the lightest atoms—carbon, hydrogen, nitrogen, and oxygen—also form very strong bonds); the best of these materials are brittle if made in large chunks, yet become tough and robust when used in fibrous forms (think of a fragile glass window vs. a boat with a fiberglass hull). In precisely made APM products, strong covalent materials can fulfill the potential they have shown in the laboratory.

Because structures made of stronger, lighter materials can be made with less mass, the costs of raw materials and processing can be reduced in direct proportion. To get a quantitative sense of the possibilities, consider some of the new materials that can be made from just carbon, hydrogen, and oxygen:

1. A lightweight material, stronger than aircraft-grade aluminum alloys, yet only one-one-hundredth as dense.
2. A lightweight material as stiff as aircraft aluminum, more than one hundred times stronger, yet only one-tenth as dense.

3. A heavier material, stronger than steel, yet only one-fifth as dense.

In each of these examples, lightweight, carbon-based materials provide strength and stiffness. In strength, materials (1) and (2) greatly outperform aluminum, enabling aerospace vehicles that greatly outperform any produced today, along with better automobiles and bicycles. Material (3) has properties more directly comparable to familiar structural metals (though only as stiff as oak), being stronger and lighter by only moderate factors. Like (1), however, it contains very little APM-fabricated material per unit volume; indeed, (3) is the same as material (1), but with void spaces filled with a hydrogel, producing a material that's more than 90 percent water by weight.

The extraordinary lightness and strength of materials like these has far-reaching consequences. Structural materials are ubiquitous. Not only bridges and aircraft, but also machine parts, tables, walls, circuit boards, and textile fibers, all consist mostly of structural materials. Better materials enable designs that offer greater strength and lower mass simultaneously. These properties can often be leveraged to increase safety and energy efficiency. Even better: These materials are also resistant to corrosion, fatigue, creep, and fracture and can be made as transparent as glass.

In aerospace technologies and a range of other spheres of engineering, the prospect of removing more than 80 percent of the structural mass promises a revolution in performance and efficiency. Performance aside, making products less massive reduces materials consumption, with a proportional favorable impact on production costs. All of these advantages follow from little more than access to stronger and lighter materials. APM-based production can, of course, do much more.

Performance Through Special Materials

APM-based methods will open new horizons in the scope of materials synthesis. Materials are characterized by their patterns of atoms, and atomically precise control can produce patterns of atoms that can't be made by today's less-controlled means. Scientists and engineers can already transform simple chemical inputs into an astounding range of materials, each with a distinctive pattern of atoms, and they perform these feats without guiding the motion of atoms and molecules, relying instead on patterns that form spontaneously.

Guiding molecular encounters can expand the range of accessible materials, and immensely. The greatest surprises will likely come not through improved mechanical properties, but through improved or novel electronic properties. Some potential applications of electronic materials and devices can be understood by considering potential applications of those already demonstrated on a laboratory scale; others can be understood through reliable calculations; while others are speculative and outside the scope of conservative exploratory engineering. These include a range of potential applications of phenomena now encountered only in physics laboratories (or theoretical papers in journals), such as coupled spin systems that may replace transistors in building digital logic, other spin systems that enable robust, topological quantum computing, new classes of high-temperature superconductors, perhaps, and a host of correlated-electron phenomena that are expected to occur in atomically precise structures that have so far eluded synthesis.

Two electronic examples play a role in the theme of radical abundance: improved photovoltaic cells for solar energy, and carbon nanotubes with walls just one atom thick that can outperform and replace scarce copper as an electrical conductor.

Performance Through Scaling

Size in itself, along with atomic precision, can offer further advantages. We've already seen how scaling can greatly increase manufacturing throughput, and the same scaling principle can apply when the product is energy. High throughput, whether of electrons or fuel molecules, can boost power density by factors of a million or more.

Today's electric motors generate force using electromagnets, but as noted before, electromagnets don't scale according to mechanical rules, and their unfavorable scaling laws preclude nanoscale motors that would work by means of their forces. Electric motors, however, can work with electric instead of magnetic fields by harnessing the flow of electric current driven "downhill" by an electric field much as water wheels harness the flow of water driven by gravity. This kind of motor works best with very small parts, on a scale where a few volts are enough to produce enormous electric fields, while the throughput of charge per unit mass can be enormous. High-efficiency systems based on devices like these can deliver kilowatts of mechanical power from an array of motors within a square-centimeter, micron-thick volume. (The outputs of many motors can also be combined to drive a large shaft, but cooling constraints then limit system-level power density.) This hugely exceeds the power density of today's best electromagnetic motors. The same kind of devices (indeed, identical devices, run in reverse) can work as electrical generators. In either direction, energy conversion efficiencies can be excellent, with losses to resistance and friction well under 0.1 percent.

Fuel cells convert chemical to electrical energy, and their performance depends on the interactions of molecules with nanoscale structures. Precisely engineered electrocatalytic and ion-transport mechanisms can boost efficiency, while mechanical scaling laws enable systems with nanoscale components to convert chemical to electrical energy with high power density.

Nanomechanical systems can also enable direct and efficient interconversion of chemical and mechanical energy; this requires a series of downhill steps in molecular potential energy coupled tightly to mechanical motions. Scaling laws once again enable extremely high power densities, comparable to those of nanoscale electric motors. An integrated system that combines chemical energy conversion with nanoscale motors and generators can serve as a zero-emission device that operates like a rechargeable battery with the energy storage density of a tank of gasoline. Systems like these could power vehicles and could be used to smooth fluctuations in solar electric power production across day and night cycles—or even entire seasons.

Remarkably, the power of mechanical scaling laws is enough to enable nanomechanical computers to compete with electronic devices (at least of the kind used today). Clock rates can reach the gigahertz range, while power consumption per core can drop to a fraction of a microwatt.

WHAT THIS MEANS FOR PRACTICAL APPLICATIONS

As the above discussion suggests, APM systems will enable dramatic advances at the most basic level of engineering, materials and devices. But leaving aside the nuts and bolts of materials, efficiency, and throughput, what does this imply for the products used in everyday life?

Look around your office, home, or street. In light of potential improvements in structures, computing, engines, and so on, almost everything you see (appliances, cars, video screens, lamps, and so on) can be upgraded in one way or another. Where measures like strength, power, energy efficiency, computational capacity, and weight are important, there's room for either improvement or revolutionary advances. Resource conservation naturally follows from lighter and more efficient products, through both more frugal use of resources in production and greater efficiency in use. With lower-cost, higher-performance heat pumps and insulation, even the energy costs of domestic heating and cooling can be greatly reduced.

Today, low-cost products often prove to be cheap, flimsy junk. Seeing this stuff replaced by sturdy, durable products, with buttons that don't break, parts that don't crack—excellent by every measure, yet low in cost—would be a refreshing change.

In solar photovoltaics today, best-in-lab conversion efficiencies now exceed 40 percent, but these cells require the use of scarce materials like gallium and indium. Photovoltaic cells made with Earth-abundant materials (such as the iron and sulfur of pyrite) promise comparable performance, but only if prepared as multi-junction stacks of thin, nearly perfect crystalline films that are then integrated with additional layers that provide electrical contacts, oxygen and moisture barriers, and anti-reflection coatings. The associated fabrication difficulties place devices like these beyond reach today.

APM can solve these problems by enabling low-cost production of atomically precise structures that consist of thin films of common materials, just tens of grams per square meter. Photovoltaics are becoming competitive sources of power today, but their use is limited by production cost and, if employed as a primary power source, by the high cost of storing energy for night-time use and cloudy weather. A substantial improvement in solar conversion efficiency, a dramatic reduction in cost, and efficient, low-cost ways of converting electrical energy into fuels and back again—together, these have the potential to change the energy economy as a whole, lowering the costs of electric power and fuel while moving toward zero net carbon emissions.

Computational systems today rely on devices that have already reached the scale of nanometers. Even so, there's still a lot of room at the bottom.

With suitable post-transistor devices, another ten-fold improvement in linear scale can be achieved, relative to today's devices, and this translates into a factor of one thousand in device density (volumetric density is the right metric to use, since APM systems can build in three dimensions). Combine this reduction in scale with device efficiencies closer to the thermodynamic limits of computation, and the prospects include micron-scale computer processor cores, memory storage densities in the range of a billion gigabytes per cubic centimeter, and billion-core computer systems suitable for air-cooled laptop computers.

Living systems are based on AP molecular components, and APM-level tools and instruments will enable medical researchers to study biological systems at that level in unprecedented detail and to then intervene with unprecedented control. The key will be fast, thorough data collection and the means for rapid development of nanoscale devices of the sort that researchers already want and are developing—for example, nanoscale packages that display biologically active, compatible surfaces, able to bind to cells of specific types, to induce endo- and transcytosis to reach difficult targets inside cells and beyond the blood-brain barrier, and to then intervene by delivering (for example) RNA and regulatory molecules that modify metabolism and gene expression. (Exosomes provide a model for such devices and have spawned a burgeoning field of nanotechnology-linked biomedicine.) Destroying emerging and drug-resistant pathogens are among the vital and accessible goals.

THE ORIGINS OF RADICAL COST REDUCTIONS

What would be the impact of a large-scale shift from traditional manufacturing to APM? The chief physical costs of APM-based production will be the costs of energy and raw materials, and these can be reduced while other costs plunge.

We live in a time when growing resource demand and scarcity are sharpening the focus on reducing raw material consumption, a time when the challenges of scarce energy resources and rising CO_2 levels have sharpened concerns about energy supply and demand while global aspirations for an escape from poverty are pushing industrial development forward with urgent economic and moral force, yet it does not seem possible to satisfy these aspirations within the constraints of current technology. How might APM help resolve these challenges?

Ultimately, cost makes the difference between the luxuries of wealth, common industrial products, and goods barely affordable in poor rural societies. Cost makes the difference between abundant solar electric power and an economy with a long-term lock-in to coal-based power. Examining APM prospects through the lens of cost is of more than narrow economic interest.

Exploring Physical Costs

In this realm, engineering methods have limited scope and can address only what can be described as "physical costs," the costs of inputs (materials, energy, labor, and land) and the harder-to-quantify costs of unwanted results, such as accidents and pollution. This accounting omits a range of costs that would appear on a producer's ledger, yet in an overall physical sense aren't costs at all, such as transfers of money through patent licensing fees, for example, or taxes (from a societal perspective, transfers of money consume no resources). This accounting also omits real costs (primarily of labor) that stem not from physical requirements, but from both organizational overhead and societal choices, the costs of management, marketing, regulatory oversight, litigation, and the like. Finally this narrow, physical concept of cost omits transaction externalities, such as the costs borne by displaced competitors (and their suppliers and workers), and the policy-dependent costs of potential disruptions in societal order and international affairs.

I will return to these more difficult questions in Chapter 16 but the first step toward a broader understanding is to focus on the next layer of physical analysis, one that follows from embedding the timeless potential of APM-level technologies in the material and economic world as we know it today. This step involves considering how physical costs can be expected to change with changes in the requirements and performance of production and products.

Costs of Raw Material Inputs

APM promises changes in the pattern of global resource demand because APM-based production calls for a different mix of raw materials. In particular, today's scarce materials will lose much of their value because Earth-abundant materials* can almost always serve the same functional purposes, and with better performance.

As for scarcer elements like zinc, tin, and lead (used, for example, to make solder, cheap metal castings, and anti-corrosion coatings for steel), it seems that most will find little use in an APM context. There will be no need for soldered joints, zinc die-castings, or tin-plated cans in a world where products can be made seamlessly and corrosion free and be of higher performance and lower in cost. Several other scarce metals (including manganese, chromium, nickel, and cobalt) are used chiefly to make steel alloys that become obsolete when steel is replaced.

What do these differences imply for the cost of raw input materials? First, the most useful elements are inexpensive. When delivered in the form of low-cost compounds (silicon dioxide, for example, rather than silicon), carbon, nitrogen, oxygen, aluminum, and silicon can be had for prices in the one dollar per kilogram range.

To understand prospective changes in the cost of raw materials, one must take account of both reductions in input costs per kilogram and

---

* Besides carbon, hydrogen, nitrogen, and oxygen, these useful abundant elements include silicon and aluminum, both common in the Earth's crust.

reductions in the kilograms required. For example, replacing steel, aluminum, and plastic structures with stronger materials would typically reduce mass by at least a factor of ten. In most instances, taking account of non-structural materials doesn't greatly change this result (and often improves it).

The bottom line: Most raw materials needed for APM products are common and inexpensive, costing less than one dollar per kilogram. With an adjustment for reductions in mass, this gives a cost equivalent to about ten cents per kilogram, a cost per "effective kilogram."

Costs of Energy Inputs

The energy costs of purification and early-stage chemical transformations (for example, in reducing silicon dioxide to silicon) may be little decreased from comparable costs today. Looking downstream and considering the energy penalty for performing highly reliable molecular fabrication steps, a reasonable overall estimate amounts to roughly one dollar per kilogram at today's electric energy prices (without adjustment for mass reductions).

The bottom line: The estimated performance-adjusted costs for energy and raw materials are roughly equal, adding energy costs of about ten cents per effective kilogram.

Costs of Land

The natural scale for a production facility is comparable to the scale of its largest products—several times the length and width would be ample. Desktop-scale systems can build desktop computers; garage-scale systems can build automobiles.

The bottom line: By comparison to current industrial systems, the land area required for APM-based production is negligible.

Costs of Directly Involved Labor

APM systems transform feedstocks into products without intervention by human hands, just as digital computation doesn't require people flipping abacus beads or writing numbers on paper.

The bottom line: The labor required for APM is external to the production process, with no labor cost incurred by the process itself.

Costs of Wastes and Pollution

By its very nature, atomically precise processing can control the structures of both products and by-products. With no need for combustion, there is little reason to produce or emit CO_2, and with precise process control there is little reason to dilute and dump wastes in the air or in water; toxic impurities removed from raw materials can be processed and returned in low-hazard forms (for example, as pellets of naturally occurring minerals that are known to be stable across geological time).

Waste heat is substantial in early processing stages (produced on a chemical-process scale, and in proportion to mass), but waste heat will have little adverse impact if removed by air cooling.

The bottom line: Advanced production systems need not produce noxious emissions, greenhouse gases, or toxic wastes, and could meet zero-tolerance emission regulations at little cost.

Risks of Accidents

The end-to-end architecture described in the previous chapter has little scope for accidents. A factory that has no room for people can't injure workers inside. A factory that contains no substantial stores of toxic, combustible, or explosive materials presents none of the usual risks to neighbors. It seems that only inventories of feedstock materials might present a substantial hazard, yet there's little reason to provide materials in forms that might not be safe. For example, there are advantages to using hydrocarbons or other organic liquids as feedstocks, but there's no need for these to be poorly packaged or volatile (to supply hydrocarbons, a fluid like mineral oil would do).

The bottom line: APM-based production systems would naturally tend to be safer than current industrial plants, and stringent safety regulations could be easily met.

Costs of Physical Capital

The physical costs of physical capital can be estimated by considering two basic features of APM: high throughput and wide product range.

Consider a full, end-to-end APM-level supply chain, from raw materials to products. The earlier throughput estimates suggest that the required machines (setting aside transportation delays) could produce a mass of products equaling the total collective mass of machinery in a time measured in hours, not days or years. Further, as is the case for today's industrial system (planet-wide, considered *in toto*), the range of accessible products includes the full range of machinery needed to build an end-to-end supply chain. If the cost of a high-throughput production system like this were to be amortized over one thousand days (quickly, by current accounting standards), then the physical capital cost per kilogram of product would be less than one-one-thousandth the cost per kilogram of the APM system itself. Since the machines themselves would be APM products, the amortized physical cost of this capital equipment can be very low indeed—unwinding the estimates above, the cost amounts to a fraction of a cent per kilogram of product. (Note that this is a marginal cost, but initial costs—R&D, for example—can likewise be amortized over a potentially enormous product base in a short time.)

The bottom line: In physical terms, the cost of production equipment adds an almost negligible increment to product cost. Indeed, the physical cost of capital per unit output would remain affordable even for equipment used at 1 percent of capacity, like a home washing machine used for just a few hours a week.

Adding Up Costs

Summing the two major costs above—raw materials and energy—while absorbing the smaller costs into the large rounding error yields a typical, estimated, physical cost of about twenty cents per effective, structural mass–adjusted kilogram. The materials and energy costs are based on conservative estimates of both per-unit costs and input requirements (conservative in part because they neglect prospects for APM-enabled reductions in the costs of the inputs).

Beyond Physical Costs: The Digital Media Model

The usual concept of production cost includes not only direct, physical costs, but also the costs of design, management, finance, marketing, sales, office support, legal services, and similar activities. Looking downstream from production, a comprehensive concept of cost must include costs of product distribution, installation, and technical support, among others.

APM-based production promises to shift these costs upward in percentage terms, but downward overall. Again, the best current model is software and digital media. APM can be regarded as a digital medium like digital sound, displays, or printing. These digital media differ, however, in the scale of required resources per unit of consumption. At the margin, resource costs are negligible for the energy required to play a song or display an image, and small for the paper and ink required to print a page, but can be substantial for the materials and energy required to produce an APM product, even at a cost as low as tens of cents per effective kilogram.

As with today's digital media systems, the primary difference between one product and the next is a digital description, like a text, audio, or video file, but now describing patterns of matter, not characters, sound samples, or pixels. And as with today's digital media, the physical cost of distributing a new product (which is to say, what is new about a product) is the cost of downloading a file.

And as with digital media downloads, the costs of management, finance, marketing, sales, office support, legal services and the like are effectively zero on a marginal, per-unit basis. Thus, today's digital media and APM systems share a similar cost structure. The cost of producing the first instance of a product is like the cost of producing a new image, video, or software application, while the costs of product distribution and delivery can be low and decentralized, and are only incurred on demand. As for on-demand publishing today, inventory is not a concern. Consumers can choose, not just from what is in stock, but from a potentially unlimited library.

Because of design costs and licensing fees (copyrights, patents), digital products are sometimes expensive to buy, and for the same reasons, per-unit purchase costs for APM products could likewise be much more than their cost of production. Nonetheless, artists, photographers, and software developers sometimes release products for free, and this trend continues in the emerging 3D printing community. The open-source model can apply equally well to APM.

PERFORMANCE, COSTS, AND PRESSURES FOR CHANGE

This chapter marks a threshold in a journey of ideas that leads from physical principles to prospects with human consequences on a global scale. The path has led from the concept of a timeless landscape of technological potential, through physics, engineering, and the methodology of exploratory engineering, to a partial view of the realm of atomically precise manufacturing and its potential products.

The next stage of this journey turns to potential applications of APM-level production that will tend to drive change in particular directions—not leading to predictable outcomes, yet shifting the balance of forces that drive human actions and affect their results. For example, if products combine lower cost with higher performance, they will tend to drive out their competitors. If a technology offers potentially decisive military advantages, then competition will tend to spur efforts to pursue or restrict it. If a technology can be applied to solve a problem with lower cost and less need for negotiation, then there will be a greater tendency toward effective action.

As we've seen, APM promises lower cost and higher performance across a broad spectrum of industrial products and processes. Potentially decisive military advantages follow, and this potential will demand a response that includes a profound reassessment of national interests.

Today's industrial civilization has set human material development on a collision course with the limits of Earth's resources and climatic stability. APM-based technologies can avert this collision, enabling effective action on problems that otherwise might be intractable.

I will return to these topics in later chapters, first exploring technological opportunities that will tend to drive change, then exploring some of the questions raised by likely responses. These questions, however, are set firmly in an historical context that has shaped today's situation and prospects. This story of progress, prospects, and conflict will set the stage for what follows.

PART 5

THE TRAJECTORY OF TECHNOLOGY

CHAPTER 12

Today's Technologies of Atomic Precision

HUMAN TECHNOLOGY EVENTUALLY LED to machines, yet it began with wood, hide, stones, and hands—which is to say, with biopolymers (cellulose, collagen) and harder, inorganic materials used to make hand-held tools. AP nanotechnologies are following a similar path, but with AP control of biopolymers and inorganic materials using assembly driven by Brownian motion rather than hands.

Once again, advances will lead to machines for making things. Just as an early blacksmith's hammer and tongs differ from an automated machine in a watch-making factory, today's early tools differ greatly from advanced APM systems. And just as blacksmith-level technology led to today's machines, so today's AP molecular technologies will lead to tomorrow's nanomachines.

Where do we stand today on the road to advanced atomically precise fabrication, the road that leads to APM? We're further along than most people think, and surprisingly so. To understand the prospects for radical abundance, we must understand the nature of today's rapidly advancing technologies of atomically precise fabrication. These don't look much like APM, yet they are paving the road.

The idea of atomically precise fabrication sometimes seems futuristic, yet this perception is far from reality. The molecular sciences began with atomically precise fabrication, and scientists gained their first understanding of precise patterns of atoms and bonds not by seeing molecules, but by learning to make them. Atomically precise fabrication began over a century ago. It's come a long, long way since then.

However, where paths to APM are concerned, this progress, though critical, has gone largely unrecognized. In fact, if not for an historical accident, today's atomically precise fabrication technologies would be seen as the crown jewels of modern nanotechnology, celebrated, better funded, and much in the news. Instead, because the relevant fields were already long established in 1986 and carried different names, these fields weren't called nanotechnology, and hence didn't share in the futuristic glamour and hype that launched and funded nanotechnology programs of a distinctly different kind.

In short, both the money and attention went elsewhere and enormous progress in atomically precise fabrication largely slipped under the radar. People looked for progress in the wrong places, saw little, and got the idea that progress had stalled. A history of ideas and politics helps to explain the gap between the initial vision of advanced, APM-based nanotechnology and the state of today's perceptions, objectives, omissions, and challenges. The following chapter will tell the story of what happened, and why.

Meanwhile researchers in the most relevant fields—today's masters of atomically precise fabrication—had little cultural inclination or immediate reason to explore their work's latent potential for systems engineering. Indeed, because their fields took shape long before the concept of APM first appeared, these researchers were fully engaged with other objectives, pursuing questions that centered on biology, medicine, materials, and small-molecule chemistry.

Let's take a closer look at these fields to explore a different question: How can advances in atomically precise fabrication enable progress in molecular systems engineering, the basis for the technologies that will open the door to APM? The critical fields have names like "organic synthesis," "protein engineering," and "computational chemistry," among others.

AP ENGINEERING WITH CHEMISTRY

Chemists make compounds/molecules, the objects of their own contemplation. That puts them close to art and artists. Lest we get too romantic about that (only someone who has not tried to make a living as an artist would), the centrality also puts us close to engineers.
—*Roald Hoffmann*

These days chemistry has a stronger reputation for stench than it does for atomic precision. In high schools, chemistry is usually introduced as a science having something to do with test tubes, safety goggles, reaction stoichiometries, the periodic table, moles of substances, acids and bases, solvents, and the like. Chemistry taught this way has little to say about the structure of molecules beyond their composition and patterns of bonds. Though teachers may broaden curricula, the traditional topics don't suggest that chemistry embraces atomically precise fabrication of macromolecules, i.e., nanoscale objects with form and function. Since few people study chemistry beyond an introductory level, most people are primed to underestimate what chemistry can do.

Across more than a century of AP fabrication, organic chemists have honed their skills in two major ways: producing molecules identical to those found in nature (pre-existing products of biological machines), and producing molecules that are highly unnatural, with novel patterns of atoms and bonding, sometimes patterns that had seemed to be impossible to make—contorted patterns of bonds, configurations with perverse instabilities.

Indeed, as Roald Hoffmann's remark suggests, organic chemists sometimes see themselves as poised between science and engineering (and the intuition and esthetics that their field demands does indeed give it an affinity to art).

Our earlier model of the contrasts between scientific inquiry and engineering design highlights a conceptual tension within the field of organic chemistry. Although organic chemists design and build things, they usually frame their field as a science. Their field is more than that, however, and in terms of the distinctions drawn in Chapter 8, the field as a whole is better conceived as an intimate mixture of science-intensive engineering and engineering-intensive science. Framing their field as a science, however, tends to focus chemists' attention in the directions just mentioned—on natural molecules (puzzles from nature) or on highly unnatural molecules that push the boundaries of physical limits. In honing their methods this way, however, organic chemists have discovered practical, reliable tools that lend themselves to an engineering approach, and extensive chemical research has pursued exactly this aim.

In a landmark 2001 paper, Barry Sharpless called for chemistry to shift its attention toward using its most reliable tools—"a few good reactions"—avoiding difficult steps, peculiar molecules, and efforts to duplicate natural structures, and instead to focus on problems defined by molecular function. Chemists had often worked this way, of course, but Sharpless was the first to clearly articulate what this goal means in practical terms, setting out criteria and metrics and giving examples.* This is, in essence, an engineering approach: Put the purpose first, then seek structures that serve the purpose while also being economical to produce.

It's difficult to properly convey the pattern of challenges and capabilities that characterizes modern organic synthesis. Studying a pile of books and reading years of journal publications can give one a good idea of the pattern, and the literature describes sets of powerful tools (click chemistries, for example), yet a conservative exploratory engineering approach entails assuming little beyond what's been demonstrated while not being too surprised when chemists deliver yet more.

---

* Sharpless called this approach "click chemistry"; one of the reactions he highlighted (and improved) works extraordinarily well and is now often called "*the* click reaction." Unfortunately, this use of Sharpless's term tends to obscure his deeper engineering insight.

The scale of this research enterprise is vast. The Chemical Abstracts Service of the American Chemical Society maintains a registry that includes (at the moment) descriptions of 67,883,986 "commercially available chemicals" (many are themselves used in synthesis), and 56,703,135 descriptions of synthetic procedures. CAS reports that the registry is currently growing at a rate of more than ten per minute.

In their quest to duplicate products of nature, chemists have developed versatile capabilities for atomically precise macromolecular engineering. In particular, chemists have learned how to synthesize biopolymers, modular molecules that consist of chains of interchangeable monomeric components, the four nucleobases of DNA, for example, and the twenty amino acids of the peptide polymers. Synthetic DNA has become a key technology in biology and biotechnology, while peptides find use not only in biological studies, but also in medicine, where some peptide drugs are now produced by the ton.

Synthetic macromolecules like these can be made to order, with sequences tailored for particular functions, encoding parts of genes and binding to biological receptors. Chemists, however, can do even more, synthesizing chains containing monomers not found in nature, thus expanding the range of functional products at our disposal.

Amide bonds like those found in nylon are used to link peptide monomers together into chains, and similar chemistry can be used to link other building blocks; for example, beta amino acids, with an additional carbon atom in their backbones; peptoids, made by an extraordinarily convenient method that puts side-chains on backbone nitrogen atoms (rather than carbon); "peptide nucleic acids" which bind DNA strands more tightly than DNA binds to itself; and others, together with all of the above in mix-and-match combinations.

Molecules of this sort, called foldamers, can be engineered to imitate biology in a crucial way. They can bind and fold to make particular structures, functional macromolecular objects that can serve as components in engineering of larger molecular systems. In foldamers, biology has served as a model for chemists, as it has done since the beginning of the field that still bears the name "organic synthesis."

AP ENGINEERING WITH
NATURE'S TOOLS AND MODELS

Genetic engineering manipulates DNA to make genes that make proteins, and protein engineering uses genetic engineering to make proteins by design. Structural DNA engineering, however, uses DNA directly.

In biology, specific proteins (together with the molecules that they bind) operate as the molecular machinery that processes molecules in digestion, metabolism, DNA copying, and protein production. Proteins form the cytoskeletons that give your cells shape, along with molecular machines made of actin and myosin that move your cells and work in muscle cells to enable you to move when the protein-based molecular machinery of neurons delivers the signal to contract.

What set me on the path to understanding the potential of atomically precise manufacturing was the idea that molecular machines of these kinds could be engineered. This was a novel idea at the time, because protein engineering was mistakenly thought to be out of reach.

Protein engineering began with the realization that science and engineering aren't the same, that the task of designing a protein that folds as intended is fundamentally different from the task of predicting how a natural protein will fold. Protein scientists developed algorithms for design, and then tested and refined them through science-intensive work that has developed a range of molecular-level engineering methodologies.

As is often true in the molecular world, the researchers developing engineering methods are scientists by background, affiliations, and culture. As a consequence, protein engineering has tended to center on scientific objectives, designing proteins in order to probe their behavior and gain a better understanding of biology.

Today, researchers routinely make proteins that fold as designed as well as proteins that bind other structures, such as metal ions, small molecules, and macromolecules that include other proteins and DNA. Researchers sometimes design proteins from scratch, following natural examples in a general way, yet unlike specific molecules found in nature. Other areas of protein research, however, follow biology more directly, starting with natural proteins and reengineering them only in part—

altering binding, pasting parts of different proteins together, and tweaking structures to improve stability.

Many natural proteins are delicate; they quickly unfold and degrade unless kept cold or frozen. Engineered proteins, however, can be far more robust. Natural proteins wouldn't last long in the heat and detergent environment of a washing machine, which is why the enzyme detergents in supermarkets contain redesigned molecules, products of protein engineering.

Seen from outside the biomolecular sciences, the potential of protein-based devices is far from obvious. "Protein" suggests "meat," and "meat" suggests something to cook and eat, a culinary rather than an engineering concern. In reality, meat is mostly water—a sort of fibrous gelatin—while proteins like the keratin in hoofs and horns are one million times stiffer than meat, and another protein (silk) has been used to make bulletproof vests.

Of course, using biomolecular machinery would seem to sharply limit the scope of foldamer engineering, allowing fabrication only of biopolymers such as the polypeptide chains that ribosomes build from the twenty genetically encoded amino acids. In reality, this apparent limit has been broken. The genetic code uses three nucleotides per codon, and each codon specifies a single amino acid. Each nucleotide has one of four bases, allowing sixty-four codons for only twenty amino acids (and stop signals). Feeding a ribosome a foreign amino acid, tagged to match (and override) a stop signal, works smoothly. In this way, a hundred or more alternative, non-natural amino acids (some very different from any in nature) have been inserted into proteins. In fact, ribosomes turn out to be more flexibly useful than one might imagine. Researchers have persuaded ribosomes to read a code with four—not three—bases per codon, allowing up to 256 alternatives.

Allowing more choices enables the use of components with special chemical, mechanical, or optical properties, and can provide parts with new shapes and binding propensities to allow for engineering of more solid, predictable structures.

Among the advantages of fully artificial foldamers, however, is that chemical synthesis bypasses all the complexity of reengineering biology and provides a direct and radical way of expanding choices. I mentioned peptoids as an attractive class of foldamers, and peptoids can be assembled from building blocks drawn from a common class of chemicals—primary amines. These provide readily accessible choices that extend into the thousands and more, and peptoids have been used to build protein-like macromolecular objects.

Starting in the 1980s, Nadrian Seeman developed structures based on branched DNA assemblies, and structural DNA nanotechnology (as the field is now called) made major strides in the 1990s. Since then, the field has gone through a series of revolutions culminating in the ability to engineer atomically precise molecular frameworks on a scale of millions of atoms and hundreds of nanometers, using a technique called "DNA origami." The simple, pairwise matching of DNA strands to form the predictable, rod-like structure of the DNA double helix has provided a molecular engineering method that can be as predictable as carpentry. In this analogy, short, single DNA strands play the role of the nails (technically termed "staples"), crossing between double helical strands to fasten them together.

The products made to date include rectangles with arrays of hundreds of DNA-addressable binding sites, struts forming octahedra, and boxes with lids that latch and unlatch. In the ten million–fold magnified view introduced in Chapter 5, a typical protein can be the size of a small child's fist, while a typical DNA origami structure would be the size of a table top.

Proteins bind proteins, DNA binds DNA, and these materials can be combined. Biology has done all the basic R&D for making proteins (so-called zinc fingers and TAL effectors) that target and bind specific DNA sequences. Researchers have begun to explore the potential of composite systems that combine these, and other, materials.

MATERIALS PROCESSING METHODS THAT ARE SOMETIMES AP

Materials processing methods can sometimes produce AP products of unique and valuable kinds, but these methods allow limited scope for design. Where an organic chemist or a protein engineer can design and build atomically precise structures by piecing together molecular building blocks, materials processing methods typically work by (for example) mixing, heating, grinding, vaporizing, condensing, dissolving, precipitating, and pounding materials, but without piece-by-piece design and fabrication, the results are hard to control. Methods like these are sometimes called "shake and bake"; success is largely a matter of luck, and the results are seldom atomically precise.

Carbon nanotubes, for example, are nanoscale tubes that can condense spontaneously from carbon vapor at incandescent temperatures (though catalysts help), while the semiconductor particles called "quantum dots" are commonly made by precipitation from chemical precursors in solution. There's no way to use methods like these to craft a range of structures with precisely designed variations.

Perhaps surprisingly, nanoparticles and nanomaterials made by shake-and-bake methods now constitute most of what is called nanotechnology, and this is one reason why today's nanotechnology has so little in common with nanomachines, atomically precise manufacturing, or a recognizable technological revolution. Indeed, a typical nanoparticle in our standard magnified view would resemble, not a machine, but a chunk of rock, while a small sample of a typical nanomaterial might resemble a billion tons of rock embedded in concrete—to qualify as nanotechnology, however, the rocks must be less than a meter in diameter. (The following chapter will tell the story of how the production of small particles and fine-grained materials came to be called "nanotechnology.")

BUILDING AP PATTERNS ON CRYSTALS WITH PROBES

Physicists in the field of surface science have learned to assemble complex, two-dimensional, atomically precise structures by using scanning probe instruments to move atoms and molecules on a surface. Don Eigler first demonstrated this method in 1990 when he nudged thirty-five xenon atoms across an atomically flat crystalline surface to form three capital letters: "IBM." These scanning-probe methods use machines to move atoms or molecules and can produce arrangements that exploit the atomic precision of a crystal surface. In this they resemble APM technologies and can be considered a partial proof of principle.

Nonetheless, scanning probe methods of moving molecules differ greatly from anything useful in APM. The tools themselves are large and seldom atomically precise. Returning to our ten million times magnified view, imagine standing on a surface in a typical experiment. Looking down at the workspace area, you'd see tens to hundreds of millimeter-scale molecular bumps arrayed across a region a few meters wide. These bumps form a pattern made atomically precise by the way the molecules spontaneously align with the atomically precise corrugations of the crystalline surface. Looking beyond this area, you'd see bumps randomly scattered over a surface that stretches far into the distance across a landscape as large as a city. Moving over the surface nearby, you'd see the scanning probe tip, a cone that tapers to an irregular point as blunt as your thumb. Looking up, you'd see the cone tower above you like a fat stalactite reaching as high as a skyscraper, and above that, the mechanism that moves the tip, a mass the size of a mountain. And with the corresponding ten million–fold scaling in time, watching the first hours-long effort to arrange thirty-five atoms would have been a task passed on to your descendants for centuries to come. While it is true that there's been some speedup since then, by any ordinary standard the throughput still remains infinitesimal.

All this is a far cry from an array of AP nanoscale mechanisms making components in a factory setting, where in the same magnified view a typical machine performing repetitive AP operations could fit in your hand and easily work at many cycles per second, corresponding to tens of millions of operations per second in real-world time.

Because of the limitations of scanning probe methods (slow, and currently limited to arranging atoms and molecules in two dimensions), I find it difficult to imagine attractive scanning probe–based paths that would lead to advanced AP fabrication. Solution-phase molecular self-assembly, by contrast, already enables three-dimensional AP fabrication on a scale of millions of atoms and in production lots measured in billions. This approach has great appeal. Indeed, from the start, self-assembly is the line of research that I have advocated as a path toward APM-level technologies. The comparatively meager state of the art in nudging molecules with scanning probe microscopes has been a distraction.

ENGINEERING MOLECULAR SYSTEMS

Much of the progress in AP macromolecular technologies has been driven by biomedical research because molecular devices are uniquely suited to biomolecular studies. Breakthroughs in reading DNA, for example, have lowered the cost of sequencing a genome from a billion dollars in the 1990s to a thousand dollars today (a cost still falling fast), and all current methods process DNA strands one monomer at a time using biomolecular devices adapted from nature. For another example, antibodies are protein-based devices widely used to sense other molecules, and protein engineering is often applied to adapt them for new applications. Likewise, because of its vital role in drug discovery, organic synthesis is yet another field in which biomedical applications drive progress in AP fabrication.

Although pursued for other reasons, these fields of research are building a technology platform for developments leading toward APM. Biomolecular engineering, in particular, is an AP technology that naturally supports progress in macromolecular systems engineering, regardless of its immediate purposes.

PUTTING THE PIECES TOGETHER: COMPOSITE MOLECULAR SYSTEMS

Today's AP molecular technologies have complementary strengths. Organic synthesis, for example, provides commercially available building blocks, functional molecules, and ways of modifying and combining them, tailoring wholly new structures, all with atomic precision. Aside from chain building, however, organic synthesis typically produces small molecules, less than a pencil's width in our standard magnified view. Molecules this size are too simple to direct complex self-assembly processes, but complementary technologies make up for this limitation in scale.

In terms of scale, structural DNA nanotechnology excels, building structures that, in the magnified view, are meters in size. Further, its regular, modular structure provides a model for achieving similar results by different means, further expanding the kit of components useful for building large intricate structures. DNA does have shortcomings, however, because with only four bulky nucleic-acid building blocks, there is limited scope for designing diverse, fine-grained structures.

Biology, meanwhile, demonstrates the rich variety of structures and functions that can be provided by peptide foldamers, which is to say, proteins. Molecules in this class complement both small organic molecules and DNA because they can bind to both and thus bind one to the other. Peptide foldamers thus can leverage the ability of DNA to form large scaffolds and the ability of chemistry to provide special components, combining both in a single structure.

Now diversify the technology portfolio by adding protein-like foldamers to the mix (peptoids, for example, with their endless range of potential building blocks), and include AP nanoscale structures from inorganic chemistry together with useful non-AP nanostructures that can be bound and organized by AP frameworks. The list could go on, adding technologies and filling in greater detail, but the overall picture seems clear enough. Complementary technologies available today provide a range of building blocks for complex, self-assembled nanosystems with diverse components and functions.

As for the range of potential applications, billions of dollars have been spent on nanotechnologies, including research on a wide range of small, functional structures, such as quantum dots, carbon nanotubes, metallic plasmonic components, "magic-size" metal and semiconductor clusters, among many others. Some useful components are atomically precise, others are not, but all these hold promise of being more useful if they could be organized like components on a circuit board, and AP frameworks with AP binders could provide the required equivalents of boards, sockets, and plugs. Billions of dollars of research funds have been spent in developing nanoscale components, and if this money has been at all well spent, then we should see further rewards from assembling these components to build more complex systems. I think that this says more than any list of potential applications, and in my experience, scientific audiences agree.

THE CRITICAL TECHNOLOGY:
COMPUTATIONAL TOOLS FOR DESIGN

Macroscale mechanical engineering today relies heavily on computer-aided design software—CAD software—at every stage, from drawing the overall forms of the parts to adjusting details, modeling operational stresses and strains, and (with further software to support manufacturing), planning paths for tools that cut metal and programming machines that move and assemble parts.

By this standard, computer-aided molecular design software—CAMD software—has made great progress, but still has far to go. CAMD software, commonly used in the pharmaceutical industry, builds on a strong base of computational modeling, the fruit of decades of research in computational chemistry. Fulfilling the future potential of CAMD (using the term in an inclusive sense that embraces macromolecular systems engineering), will require integrating multiple methods of modeling coupled closely with both human-guided and computational design.

Human guidance here plays its usual role, shaping the overall system architecture and designing parts that can implement it. Modeling likewise plays its standard role: Given a design, describe how it will behave. What I've called "computational design," however, is different in the macromolecular domain: The parts of irregular foldamers must fit together like a three-dimensional puzzle, but the specific set of parts isn't given first and the best fit is never precise and obvious. Software instead takes a general description of molecular form and function and searches a vast space of possibilities for monomer sequences that will fit together when folded and perform their intended function.

At the moment this kind of software exists only for proteins, though it will soon be extended to peptoids. The automated computational design search processes must further be integrated with system-level role definitions and with the design of DNA structures, bound organic molecules, description, and so on. There's no great mystery about how to do this (part of the work is already done), but success will require a lot of hard work along with an open architecture that enables contributions by many loosely coordinated groups.

CAMD capabilities are coming together today in a patchy fashion, but not yet with the integration, systems-level focus, and the breadth of coverage that the effort deserves. The pace of improvement in CAMD software will likely set the pace of progress in systems development—fabrication technologies are already available, courtesy of chemists and biotechnologists, which leaves design as the primary problem.

ROADS TO REVOLUTION

A broad, open road leads from current technologies to APM. We've come a long way, and there's a long way to go, but the road has no gaps, no chasms to cross. The road leads from current AP self-assembly technologies across a landscape of molecular technologies for building AP machines. Here, I will sketch the lay of the land and the directions that lead upward toward advanced capabilities.

Two Methods of Assembling Parts

Advances in stereotactic control—the ability to direct molecular assembly by guiding molecular motions—will open new horizons; researchers today have gotten no more than a taste of its power. Stereotactic assembly is how you put a peg in a slot, or a shelf into a bookcase; it's just the generic, common-sense way of building things: *Move parts into place and put them together.* The self-assembly method, by contrast, seems strange: *Let parts move at random and require each part to encounter and bind to its own special place.*

Researchers learned the principles of self-assembly from molecular biology and now use it extensively; in the molecular sciences, self-assembly seems like common sense, while putting parts together with direct control often still seems strange.

To understand the next steps and the road ahead, it's important to understand how self-assembly and stereotactic methods can be combined in complementary ways.

Today's Self-Assembly Methods

Self-assembly has one great advantage: Because it employs thermal motion to move parts into place, assembling parts requires no nanoscale machinery. Researchers can use conventional biological or chemical means to make the parts, and if they're properly designed, thermally driven Brownian motion can do the rest.

Self-assembly, however, also brings with it a set of challenges and limitations. To produce complex, non-repetitive structures by self-assembly, each part must bind in just one specific place and position as the structure comes together, and this means that each part must have a unique shape, like a piece of a jigsaw puzzle. These molecular puzzle pieces can't be too small, or they'd be too simple and too much alike; they can't bind too strongly, or near-matches would end up in the wrong places and never let go; and like a finished jigsaw puzzle, the end product would be divided by many irregular seams.

The overall challenges of self-assembly make design difficult, and today, folded, polymeric macromolecules are the only structures large and complex enough to form suitable components. These challenges can be met, however, and as we've seen, self-assembled macromolecular products can be useful.

Advanced Stereotactic Methods

The strengths of self-assembly and stereotactic control can be combined. To see how, first consider the power and challenges of advanced stereotactic control.

At a sufficiently advanced level, stereotactic control can be used to implement APM, and as described in Chapter 10, the most advanced forms of stereotactic control require moving small reactive molecules along complex, tightly constrained paths. The motions must be complex because the machinery must guide, position, and orient many different parts, and the motions must be tightly constrained because the parts must not be allowed to miss their targets, despite thermal fluctuations. Exploratory engineering shows that this requirement can be met by means of complex machines built of rigid parts that, in turn, are made of fine-grained, tightly bonded materials of high elastic modulus. To construct such machines, however, will itself require high-quality stereo-tactic control. This may seem circular.

At first glance, then, APM-level technology might seem like a castle in the air, attractive, perhaps, but with no way to build it. Indeed, this first-glance impression has been quite common, because enthusiasts for advanced nanotechnologies have envisioned the highest imaginable tur-rets, while researchers in the molecular sciences have been too preoccu-pied to envision plans for building foundations.

Practical efforts must start at today's level (already far above what once was the ground), and then build higher levels, layer by layer, with blocks made of available materials. Identifying suitable architectures is a task for exploratory engineering that addresses a series of fabrication con-straints; each level of capabilities—each emerging class of systems—must be implemented using capabilities made available by the level below. The key to understanding the next several levels is to see how self-assembly and stereotactic control can mesh to form a gapless continuum.

Putting the Two Methods Together

The principles of self-assembly and stereotactic control work well to-gether. Stereotactic control can repair a weakness of self-assembly, and self-assembly can reduce the challenges of stereotactic control. Loose stereotactic control can easily constrain general positions, and self-assembly then can provide fine-scale alignment. Systems that combine both methods this way can be relatively simple and soft, and need not look anything like an APM system.

These complementary strengths can be combined to provide a pow-erful approach to AP fabrication:

- Use coarse-grained building blocks to build structures with well-separated, easily targeted binding sites.
- Use soft machines to select specific sites, allowing enough freedom of movement to enable self-alignment.
- Use self-alignment and binding (localized self-assembly) to lock each building block into a precise position.

Compared to pure self-assembly, this approach allows the use of smaller building blocks, easier to make and with more standardized designs, yet more diverse in their properties and useful for making a wider range of products. And better yet, the required machines are themselves within reach of fabrication by pure self-assembly, through extensions of techniques already in use. With a focused, coordinated effort, protein engineering, organic synthesis, and structural DNA nanotechnology are more than enough to construct the kinds of components required; the critical challenge is design, not synthesis.

And as for using smaller, more diverse building blocks to construct a wider range of products, these products can include machines that enable tighter stereotactic control. Finer grained, more rigid materials and greater ease of design can be applied to build machines of greater rigidity and complexity, thereby enabling more precise control of more complex motions, and hence yet tighter stereotactic control. This paints a picture of a gradient of technologies that leads from current laboratory capabilities to APM-level stereotactic control, the threshold of the APM revolution.

There is indeed no gap.*

IN LIGHT OF ADVANCES in AP fabrication technology and considering the path I have outlined, why haven't we seen greater progress toward APM-level technologies after more than a decade of well-funded nanotechnology research?

The next chapter examines the history of the selling, funding, and redirection of a nanotechnology program that had promised advanced AP fabrication, and then with funding in hand, redefined nanotechnology to mean something quite different. The ends, means, and funding parted company and went in different directions.

The present state of affairs cannot be understood without understanding what happened.

---

* Appendix II surveys the technology gradient in more detail, exploring additional aspects and intermediate points.

CHAPTER 13

A Funny Thing Happened on the Way to the Future . . .

EVERY MAJOR NATION now supports nanotechnology research. Some world leaders regard nanotechnology as a crucial part of the future and share a vision linked to my work. For example, when the president of India, Dr. Abdul Kalam, who had formerly led India's missile and space program, delivered a series of addresses in which he spoke of nanotechnology as India's future, he cited both Richard Feynman's talk, "There's Plenty of Room at the Bottom" and my technical book, *Nanosystems*. In 2001 I met Dmitry Medvedev, president of Russia, after he delivered an address on nanotechnology at the opening session of Rusnanotech 2011; he had read my first book, *Engines of Creation*. More than a decade ago, in support of a vision of atomically precise fabrication, President Clinton announced a plan for the world's first national nanotechnology program; today, China's program has a larger (and rapidly growing) budget.

Today, government spending on nanotechnology research has reached a billion dollar scale in the United States, Europe, and China, and corporations fund research at a comparable level. The fruits of this research appear in scientific journals, with more papers published daily than anyone could read.

One might expect that progress in AP nanotechnologies would have rolled forward with this flood of research, and it has, but not to the extent that might have been expected. Between the initial inspiration and the rise of national programs the story took an unexpected turn.

In the United States, the leaders of the nascent program sold a vision of nanotechnology centered on atomically precise fabrication, but then redefined "nanotechnology" in a way that effectively severed this link and discarded all mention of atoms. If this seems implausible, please keep in mind how strange politics can become and how far beliefs can stray from reality. Sociologists and historians may want to review the public record (outlined below), and reassess the academic literature on this episode.

As we look toward the future, it will be important to understand how the facts became twisted, both to correct past misimpressions and to avoid renewing them. I offer this story to help explain puzzling facts about today's conversation and as a cautionary tale for the conversation to come.

The story begins with a word and its meaning.

TWO KINDS OF NANOTECHNOLOGY

Some nanotechnologies feature atomic precision and others don't. The line can be fuzzy, yet as we've seen in earlier chapters, it marks an important practical difference. As it happens, it also marks a boundary that became a political fissure, one that split nanotechnology in two and set the two parts at odds.

Nanotechnology as Atomically Precise Fabrication

Nanotechnology began with the concept of atomically precise technologies, and as we have seen, atomically precise technologies are thriving today. They include fields within chemical synthesis and biomolecular engineering along with a range of AP technologies developed by branches of physics and materials science. Previous chapters have explored some of today's AP nanotechnologies (million-atom frameworks, foldamer structures, and so on) and have outlined progress toward a new discipline of atomically precise molecular systems engineering.* These thriving AP technologies stand on their own merits, regardless of longer-term expectations, and it is their intrinsic merit that has driven progress on paths that lead toward APM.

Nanotechnology as Nanoscale Materials and Devices

As part of the redirection of the US program, nanotechnology was re-defined solely in terms of scale, eliminating all mention of atomic pre-cision. To qualify as a nanotechnology, it was merely necessary that structures have features with "dimensions of roughly one to one hundred nanometers." AP nanotechnologies often satisfy this size criterion, but so do transistors on silicon chips and particles of ultrafine powder.

Atomic precision is important, but shouldn't mark a deep divide be-tween fields. AP and non-AP nanotechnologies overlap in methods and applications, share instrumentation and modeling techniques, and in combination the two are more powerful than either alone. Incorporating AP nanodevices into non-AP silicon circuits, for example, may emerge as a leading technology in the global semiconductor industry. In fact, AP and non-AP techniques are both parts of a spectrum and will develop as partners. In fields that range from biomedicine to structural DNA nanotechnology, they already are.

Nonetheless, for reasons outside the technical realm, the relationship between these two kinds of nanotechnology was uneasy from the start, and it soon became tense.

Two Kinds of Promise

What launched the field of nanotechnology with all its excitement and prestige wasn't a sudden infatuation with nanoscale particles, or even

_____

te that I had titled my 1981 paper "Molecular Engineering: An Approach to the Development of ral Capabilities for Molecular Manipulation," and proposed pursuing exactly this line of research.

transistors. At the outset, "nanotechnology" was simply a name I had chosen to label the concept of an APM-based technology, a name that occurred to me between the first and second drafts of *Engines of Creation*.

Without the promise of APM-level technologies, nanotechnology in the broader sense would have progressed less quickly and very likely under a range of more traditional names. There would have been no abrupt takeoff of press coverage, no public fascination with a nanoscale robot mythology, and no reason for nanotechnology to infiltrate popular culture through books, movies, and computer games. Nanoscale particles and the like would never have been mistaken for a technology that could upend the world.

Instead, the fascination began with the 1986 publication of *Engines of Creation*. Feature articles and coverage in the popular press reached millions of readers within a few months. Science fiction novels took up the theme in the years that followed, further exciting the public's imagination. During this time, "nanotechnology" in the public mind grew into a vision of a futuristic technology based on tiny machines, loosely derived from my initial conception of high-throughput, atomically precise manufacturing. Mention of other nanoscale technologies was rare at the outset and contrasting concepts only slowly gained traction.

Research in nanoscale materials promised better materials and research focused on nanoscale devices promised better electronics, sensors, and the like. Developments like these can be valuable on a scale of millions or billions of dollars, yet they aren't at all like APM. The technologies are different, and the potential rewards are incommensurate. Indeed, from a societal perspective, the longer-term promise of APM-level technologies inherently overshadows any possible near-term results of research and this has been true from the beginning.

Conflation, Confusion, and Conflict

The 1990s saw increasingly widespread confusion between near-term and long-term technologies, and this confusion suggested a close relationship, a narrow gap, or a short path from present technologies to prospects that had rightly been regarded as decades in the future. Nanotechnology,

it seemed, had already arrived. The confusion, however, served to channel money to researchers who then had little incentive to explain the difference between nanoparticles and nanomachines. The incongruity led
to tensions.

Imagine the position of researchers specializing in making, studying,
and applying the properties of very small particles. In the years before
1986, their studies had little cachet, yet in the early 1990s the world increasingly found their research exciting—provided they called it nanotechnology.

Indeed, researchers from not just one, but a host of fields were rewarded
with interest when they referred to their work as nanotechnology—and
why not use this label? The word in itself fit well enough, because their
work was "nano" and also technology. As researchers followed one another in adopting this label, it began to serve a real purpose, bringing
researchers together across academic boundaries to create new communities. Their differences sometimes strained any plausible sense of relationship, yet the banner of nanotechnology still brought them together,
bringing the pleasure and rewards of long-overdue recognition. Something called "nanotechnology" seemed to explode.

In those early days, the appearance of growth stemmed primarily
from relabeling research, which became a well-known tactic for winning
funding. People joked about this at conferences and asked a question
that has never gone away—"What *is* nanotechnology, anyway?" I know
of no other field pasted together from pieces that had so little in common,
and certainly none defined by a criterion as generic as size.*

Thus, as funding followed relabeling, more and more groups crowded
into the tent and they found a shared interest in promoting and controlling the label itself. The term proved elastic and got stretched to embrace
everything from fibers to paint to pieces of metal with a fine-grained

---

* Just after writing this paragraph I picked up the current issue of *Science* and encountered an example
of how the term has stretched: a title with the words "Iron Oxide Nanoparticles" and an abstract that
begins with "Medical applications of nanotechnology . . . " and ends with a reference to " . . . using nanotechnology to activate cells." The paper describes a ground-breaking application of specific kinds of
small particles, but there is no distinct "nanotechnology" to be found in the paper, and so the use of the
term amounts to promotional branding.

crystalline texture. This kind of nanotechnology eventually drew billions of dollars of funding. A firm grip on the term had great value, and the term itself became a brand worth protecting.

Regardless of what the promoters might say, connections with APM remained vague. The idea of advanced, atomically precise fabrication (of some sort, achieved somehow, someday) remained a powerful slogan, yet became no more than a free-floating dream with no implementation plan. Confusion became both profitable and institutionalized, while the concepts that had launched the field became more a matter of rumor than study.

Meanwhile, although the rewards of calling materials research nanotechnology were compelling, there were also costs. The term "nanotechnology" brought with it both expectations that couldn't be met and fears that had no connection to anything real.

―――

IMAGINE YOURSELF as a researcher in the early 1990s, studying nanoparticles under the banner of nanotechnology. In speaking with people outside your field, you and your colleagues receive greater respect, but also encounter misplaced expectations and misplaced fears. People ask you peculiar questions like these:

"When will you build us those tiny robots we've read about, the ones that will clean out our arteries?"

And then:

"Aren't you afraid that those tiny robots you're building will run amok and destroy the world?"

The first kind of question is annoying because it sets impossibly high expectations; the second kind is disturbing because it suggests that your work is dangerous. The nanoparticles you make in the laboratory aren't at all like machines, yet excitement about nanotechnology keeps morphing into talk about tiny robots. In fact, you find that the press seems positively obsessed with the idea, and their obsession kept growing. What can you say?

These questions have nothing to do with your work, or with that of your colleagues, or with anything attainable through any current technology. In such a situation it's almost a duty for scientists to reject popular fantasies and set the record straight.

And yet, crucially, these questions often came with a third question: "When will nanotechnology give us swarms of tiny robots that can build almost anything, atom by atom?"

This was when the very idea of APM began to raise ire. The concept had become closely linked with promises and dangers that seemed (and often actually were) absurd, and atomically precise fabrication machines—which were all seen as the same—had morphed into imaginary swarms of tiny, threatening, atom-juggling robots.

The easy, uninformed response to this strange bundle of ideas was to deny that they made any sense. The promise of an AP technology revolution was still advertised, while the technologies themselves were first misunderstood, then rejected.

What had happened?

A PURPOSE, A SUCCESS, AND THE BIRTH OF A MONSTER-MEME

Some of the public's misunderstandings can be traced back to *Engines of Creation*. The book set in motion a train of events that were at first along the lines I'd intended. In particular, it introduced the public to the immense potential of advanced AP nanotechnologies, both positive and negative; it imparted a sense of the scale of potential benefits and risks alike, as a package. In this, I succeeded (again, only in part), yet there were those who called *Engines* utopian, others who described it as balanced, while some found in it the seeds of a nightmare, or pure fiction. Yet others saw the book as an initial survey of inherently mixed possibilities, a point of departure for further consideration by a forward-looking segment of the public.

As I said in describing how this story began, I have long been driven by a sense of mission in life, a mission to help solve global problems. In this, I saw my role as one of exploring potentially transformative technologies, studying them, and sharing what I learned. Having discovered the potential of advanced nanotechnology, I couldn't ignore it, and my sense of mission set my direction. My aim wasn't to promote advanced nanotechnology, but to share a balanced view of its implications—to describe potential benefits while warning of potential risks, all decades or more before those risks might become an urgent concern. Enthusiasm for the benefits of APM-level nanotechnologies seemed inevitable, and with this enthusiasm I expected a push for development. As we've seen, there was indeed a massive push that seemed to be aimed at development, but it was coupled with claims that nanotechnology had already arrived, bringing forth premature, misplaced concerns and expectations.

In considering a realistic timeline for APM-level technologies, I saw a need (not urgent, yet crucial) for a gradual growth of understanding that would provide a basis for navigating a difficult societal transition, a basis for developing regulatory mechanisms to oversee a range of potential advanced technologies, and a plan to implement what would amount to a system of arms control.

Overall, my effort to impart a sense of concern had succeeded, but the nature of the response went awry. Along with more serious concerns, I had outlined an *easily avoidable* risk that would involve small, dangerous, but useless self-replicating machines.* This vivid but secondary concern quickly diffused through popular culture and grew in the telling. Tiny machines that would somehow run amok came to be seen as an enormous, intractable risk, one that was somehow inherent to nanotechnology. If this had been the only confusion, it would have done little immediate harm and over time might have faded.

The problem was that the public saw something called "nanotechnology" arriving long before schedule, imbuing confused and distant

---

* These hypothetical machines (in swarms dubbed "gray goo") had a superficial similarity to a soon-obsolete concept for how one might organize nanoscale production machinery to build large products; this idea stemmed from a biological analogy (think of plant cells dividing while building trees), and it soon took root in the popular mind, hybridized with "gray goo," and evolved into a vivid, tenacious mythology. In the exploratory engineering world, the more practical and efficient factory model soon superseded the earlier, less practical, and more easily distorted concepts, but factories somehow weren't as exciting as swarms of nanobugs and had little traction in the press or the popular imagination.

concerns with a sense of immediacy. When researchers attached themselves to the vision of advanced nanotechnology, they found that these fears had become attached to their work. In the popular imagination, nanoparticles became kin to swarms of dangerous mechanical nanobugs—that intractable, inherent risk of the mysterious nanoworld.

And so nanotechnology researchers found themselves confronting persistent questions about nanomachines, building with atoms, and tiny robots that might run amok, all muddled together and attributed to the author of a book called *Engines of Creation*. And they attempted to squelch these questions, not by correcting misconceptions, but by wholesale rejection of a broad swath of ideas that they, too, had misunderstood.

Visions, Funding, and a Perfect Storm

The same fuzzy perceptions that enabled machines to become bugs in the public imagination also enabled a hodge-podge of nanoscale technologies—not molecular engineering—to appear to be the leading edge of the APM revolution—or something like that, though nothing was clear.

Research groups climbing on a bandwagon is nothing new, and this bandwagon effect can bring benefits for science. The new nanotechnology helped to cross-fertilize fields, gave rise to new lines of research, and brought both scientific knowledge and practical applications. And, of course, this broad range of research also contributed to technologies relevant to APM.

In retrospect, however, a clouded perception of facts marked the start of a perfect storm of dreams, nightmares, and confusion. The dreams boosted efforts to bring federal funding, while the nightmares threatened to block it, and confusion ensured misguided responses. As controversy peaked, leaders in Washington, driven by the politics of greed and fear, lashed out to destroy what they thought stood in their way.

And all this began with fuzzy perceptions, with misunderstood facts about what physical law implies for potential technologies. Indeed, because exploratory engineering wasn't within the conceptual frame of science, few scientists saw that there could *be* objective facts of the matter, and so few sought to look for them.

A Field Wedded to Ambiguity

There was a deep, structural problem with relying on ambiguity to market a field. Clarity is important in science, yet clarity about the nature of the new nanotechnology would have undercut its support, because that support, at first, had grown out of confusion. The problem was at its worst in the formative years when the promise of APM-level technologies—and radical abundance—had set the level of public expectations, while returns from research into nanomaterials (and particles, fibers, non-AP devices, etc.) had scarcely begun.

Accordingly, in the early days the new community embraced the rhetoric of "building atom by atom" because this was part of adopting the nanotechnology brand. The sticking point, however, was the idea of using nanoscale machines to build with atomic precision. This prospect was, at the time, beyond near-term research horizons and, as we have seen, it carried a lot of baggage. Nonetheless AP fabrication using AP machines remained essential to the concept, and hence beneath all the fuzz, nanomachines remained the basis for the vision of nanotechnology as a profound revolution.

The community's natural response was to retain the idea of "building atom by atom" as part of their rhetoric, while rejecting the idea of nanomachine-based production (tiny robots!) as a fantasy—as indeed it was in the popularized form that had forced itself on their attention.

It made no difference that an APM factory-in-a-box isn't a microscopic robot or that nanomechanical systems offer the only known way to deliver the promise of nanotechnology: Instead, fuzzy clouds of popularization hid almost everything. Technical concepts never penetrated the fog, and so the stories in circulation weren't about molecular systems engineering, or about factories, physics, elastic restraints on molecular motion, convergent assembly of microscale blocks, and systems-level design and analysis. Instead, they were all about swarms of nanobugs and neither facts, nor up-to-date concepts, nor technical publications could anchor discussions to reality.

———

EVENTUALLY, building on public excitement and the support of emerging research communities, a Washington-centered leadership pursued funding, and success was at hand. In 2000, President Clinton proposed and then Congress established a billion-dollar federal program, the National Nanotechnology Initiative (NNI).

It seemed that the promised nanotechnology revolution had shifted into high gear—provided that one hadn't followed developments closely and with a well-informed technical eye. And even then, no one could have guessed what would happen next. Indeed, I found that few recognized what had happened even after the fact.

At the Turning Point: A Vision Promised and Betrayed

What did the NNI promise to Congress and the public? In brief, its architects proposed to support progress toward atomically precise manufacturing, or something much like it, and Congress funded the program to do exactly that. In light of subsequent events—the NNI's quiet repudiation of its mandated purpose—it's important to remember what happened. The facts seem to have slid down a memory hole.

In 1999, the architects stepped forward to issue a glossy "brochure for the public," entitled *Nanotechnology: Shaping the World Atom by Atom,* which asked "What if we could build things the way nature does—atom by atom and molecule by molecule?" and then went on to describe the wonders of a future tied to a vision of atomically precise nanotechnologies. This lengthy vision statement was issued under the aegis of the Interagency Working Group on Nanoscience, Engineering and Technology in conjunction with the cabinet-level National Science and Technology Council (NSTC).

Then, in July 2000, the Working Group and the NSTC issued a more formal document, titled *National Nanotechnology Initiative: The Initia-*

tive and its Implementation Plan, transmitted by the White House and addressed to members of Congress. In defining nanotechnology, the *Implementation Plan* stated that "the essence of nanotechnology is the ability to work at the molecular level, atom by atom, to create large structures with fundamentally new molecular organization," and went on to speak of "gaining control of structures and devices at atomic, molecular, and supramolecular levels and to learn to efficiently manufacture and use these devices," and of "devices at the atomic/molecular scale," and of "natural nanomachines" that perform atomically precise fabrication. The vision seemed clear enough.

Congress funded the National Nanotechnology Initiative on this basis and included an explicit mandate to pursue atomically precise fabrication. In the section titled "Definitions," the bill states that "the term 'nanotechnology' means the science and technology that will enable one to understand, measure, manipulate, and manufacture at the atomic, molecular, and supramolecular levels" (note that "supramolecular" refers to molecular self-assembly, part of the natural path forward in AP molecular systems engineering).

These documents describe an APM research agenda, first promised, then funded. I wasn't consulted, but I would have approved.

Then, something remarkable happened and I know of no parallel. It's as if NASA had sold the dream of spaceflight, then turned around and rejected rocketry while still promising the Moon.

With funding in hand and a new administration in office, the architects, now leaders with significant power, purged the NNI's plans of any mention of atoms or molecules in connection with manufacturing and redefined nanotechnology to include anything sufficiently small. Tiny particles were in, atomic precision was out. Again, public documents tell the story.

Congress had drafted legislation that defined nanotechnology as a field "in which matter is manipulated at the atomic level (i.e., atom-by-atom or molecule-by-molecule)"*, echoing the promised "ability to work

---

* Much later, in 2008, a bill was introduced that would strike "atomic, molecular, and supramolecular levels" from the law, and substitute "nanoscale."

at the molecular level, atom by atom," yet the NNI then stated a different agenda, one that removed "the essence of nanotechnology" to redefine the field as follows:

> "Nanotechnology is the understanding and control of matter at dimensions of roughly 1 to 100 nanometers. . . . At this level, the physical, chemical, and biological properties of materials differ in fundamental and valuable ways from the properties of individual atoms and molecules."

In other words, NNI's nanotechnology was defined by size alone— and not in terms of atoms and molecules, but in contrast to them. To see how much the plans had changed, consider what the new document no longer mentions.

In 2000 the NSTC *Implementation Plan* for the NNI had mentioned atoms more than one hundred times in expounding a vision of atomically precise nanotechnology. Its successor document, the 2004 *National Nanotechnology Initiative Strategic Plan*, mentions atoms in only three places, and just once in its body.

In a cover letter, Presidential Science Advisor Neal Lane describes nanotechnology as "assembling materials and components atom by atom, or molecule by molecule."

In the main text, however, the sole mention appears in the definition quoted above, which places atoms and molecules outside the scope of the new nanotechnology.

Finally, atoms reemerge in the bibliography, which cited what had been promised: *Nanotechnology: Shaping the World Atom by Atom.* "Nanoscale" scores sixty mentions, but the "essence of nanotechnology" had quietly and completely disappeared.

———

IN SHORT, nanotechnology now included everything except the essence of what had been promised and funded. Progress in atomically precise fabrication continued to accelerate, but in the molecular sciences, without significant support from the NNI, and seldom called "nano."

Did the NNI fund valuable research? Yes, of course! Has some of this research advanced AP technology development? Yes, of course! Broadening nanotechnology to include more fields of research brought many returns. The problem with the NNI wasn't inclusion, but exclusion— not that it supported a wide range of research, but that it had usurped and rejected the heart of the field, absorbing widespread support for advancing "the ability to work at the molecular level, atom by atom" while instead doing nothing of the sort.

A change of this magnitude in a federal planning document is neither a casual omission nor a mere change in wording. Shoving the concept of atomically precise fabrication down the memory hole was just one part of a dirty war within science. And the NNI, now channeling billions of dollars, had very big guns.

Something strange had happened on the way to a billion-dollar program—but what? Let's rewind the story to the days when NNI funding seemed close, but threatened.

Panic and Funding in the Year 2000

For the NNI leadership to drop their mission suggests an internal upheaval beyond the normal range of politics. I can offer only a partial explanation for this sudden turn. The key, I think, is a sequence of events in mid-2000, events that emerged from the spiraling fantasy conflict of the 1990s.

In January 2000, President Clinton called for the creation of a national program to support research in nanotechnology. And then in April 2000, Bill Joy of Sun Microsystems published an article on future technologies in *Wired* magazine—a thoughtful, wide-ranging, widely noted essay—in which he raised alarms about the future of robotics, genetic engineering, and nanotechnology, and stated his concern that these technologies might, in some form, lead to the extinction of the human race. Bill Joy attributed his view of nanotechnology to me, through *Engines of Creation,* and called for a ban on nanotechnology research. Press coverage and television interviews followed, then a public response that threatened to upset politicians.

Congress was expected to vote that fall to approve or reject the proposal to create the NNI, and now it was easy to imagine that public fears would snowball, yanking away the prize.

Meanwhile, the leaders pushing for the proposal gave every sign of believing that APM-level fabrication technologies meant nanobugs, false promises, and nonsensical threats, and they seemed to think that these ideas were all of a piece and nothing more than a pernicious error. My name had already been tarred by the mythology, and now Bill Joy had inadvertently set me up for attack.

Public documents offer a glimpse of the state of mind in the leadership's inner circle. In a report from a September workshop, their *de facto* scientific spokesman, Richard Smalley,* indicated what they saw as the threat:

> The principal fear is that it may be possible to create a new life form, a self-replicating nanoscale robot, a "nanobot". . . . These nanobots are both enabling fantasy and dark nightmare in the popularized conception of nanotechnology. . . . We should not let this fuzzy-minded nightmare dream scare us away from nanotechnology. . . . The NNI should go forward.

In other words, the clamor was all about nanorobotic bugs, funding, fear, and politics, far from anything reality based; in 2001, in the pages of *Scientific American,* Smalley explicitly equated APM, in the most general sense, with swarms of dangerous nanobots (potentially intelligent and conspiratorial, no less). Around that time, in his congressional testimony and other statements, atomically precise fabrication swung back and forth between being essential and impossible while my role in the field, in his view, swung from my being the man he acknowledged as inspiring his enthusiasm for nanotechnology, to my being an ignorant fellow, beyond reach of reason, and guilty of scaring "our children" with tales of

---

* Who in 1996 had shared a Nobel Prize for the discovery of buckminsterfullerene, a beautifully symmetric molecule that forms spontaneously, for example, in candle flames.

monster nanobots that he claimed were my invention. (Smalley subsequently spoke out against Darwin.)

One can see how panic layered on top of confusion like this could lead to rash actions, and why the NNI's leaders might have tried to push hard against the ideas they perceived as a threat, yet what happened went beyond this. For leaders to embrace and sell atomically precise fabrication, then quietly expunge any mention of atomic precision—if not for the public record I would find this hard to believe, and even in retrospect I find it hard to explain.

Repression

The sharp twist displayed by the public record suggests quieter pressures, more easily hidden. Such pressures were indeed there and within the scope of NNI-led nanotechnology they enforced a kind of party line.

For example, in the late 1990s the United States had a small but thriving research program in molecular manufacturing (a term synonymous with APM), advised by colleagues of mine and based at the NASA Ames Research Center. The program funded external researchers at Caltech and Oak Ridge National Laboratory to pursue computational modeling for nanomachine engineering. With the rise of the NNI, the Ames program was quashed.

As an interagency program, the NNI had influence throughout the federal government (not only in NASA), and so it became a single point of failure for nanotechnology leadership in the United States. With a central nexus of funding (and defunding), commenting on the emperor's lack of clothes was unwise. A perception of career risk is enough to induce self-censorship, and within the NNI orbit and against the background of earlier polarization, the risk was real and stifled discussion.

Many nanotechnology-oriented researchers (not only Smalley) had been inspired by *Engines*. Some had directed their careers toward nanotechnology when the field still stood for its vision, yet now they felt pressure to say nothing about it, much less discuss, explore, formulate, and pursue coherent research programs.

The opportunity costs were incalculable. Funding patterns were distorted, of course, but more important than that, the apparent experts in nanotechnology had entrenched their own misconceptions, then exported their views.

Other countries around the world suffered a measure of collateral damage. Their scientific leaders naturally believed that the NNI was a cutting-edge program directed toward its purpose as promised and funded. In other countries, the aim of joining the global effort to develop something like APM-level technologies naturally translated into supporting domestic research that followed the NNI's lead. It would have been difficult to make any other decision, and (though without the virulence) the same dynamics of ambiguity, hype, and funding were at work in the background, a similar configuration of minds and interests.

The greatest damage, however, had been done to the United States itself. Nanotechnology funding today is nearly at parity across the United States, Europe, and China, but what this means in practice is almost as unclear as the meaning of "nanotechnology" itself. What does seem clear, however, is that US strength in many of the most relevant fields has been offset by a unique and self-inflicted wound.

These problems, of course, are temporary. Struggles fade, new leaders rise, opinions change, and actions follow. Even in the United States there's been a strong rebound from the times I've described.

When I reach outside the United States nanotechnology community, I find less commitment to error, less entrenched confusion, more open discussion and respect for facts. The rejectionist syndrome fades with the passage of time, and with distance as measured by geography, cultures, institutions, and academic disciplines. Independent studies have also helped correct the bias.

Toward Recovery: A Federal-Level Study and Roadmap

In the United States, the National Academy of Sciences serves as an independent source of scientific advice for the federal government. In 2003, the House of Representatives passed a bill directing the National Academy of Sciences to conduct a study on molecular manufacturing to "determine the key scientific and technical barriers ... examine the current state of the technology ... [and] review current and planned research activities," all as part of a triennial review of the NNI. After the House made this request, a leader from within the NNI orbit lobbied a Senate member of the House-Senate conference committee to strike the request from the bill, and succeeded. The National Academy of Sciences, however, was free to expand the scope of its studies and decided to proceed.

When issued some three years later, the National Academy of Sciences report reviewed the analysis I'd presented in *Nanosystems: Molecular Machinery, Manufacturing, and Computation* and called for funding a program of experimental research to explore APM technologies and their potential development paths. Today, anyone seeking institutional validation of APM concepts can point to the results of this study.*

In 2005, as the climate of opinion continued to improve, it became possible to take further steps. Alex Kawczak (vice president for BioProducts and Nanostructured Materials at the Battelle Memorial Institute) organized a roadmap project hosted by several of the US National Laboratories where Battelle manages research; I joined him as co-leader. The project and its two hundred participants explored directions for research toward APM. Alex told me that when the study began, potential participants were afraid to join for fear of damaging their careers, while toward the end of the study, potential participants had become eager.

With the passage of time, repression has faded, though I'm sure we'll hear echoes in the years to come.

SCIENCE, ENGINEERING, AND OPPORTUNITY

As we've seen, the missing ingredient needed for progress toward APM isn't a breakthrough in science, but a breakthrough in purpose and organization. Because the NNI seemed to fill that role while in fact impeding it, we're left with the original, structural challenge of building a new field of atomically precise engineering on foundations developed within a culture of science.

---

* To no one's surprise, the NNI made no reply to the report's recommendation.

The opportunities are greater today than ever before.

With advances in fields like structural DNA nanotechnology and protein engineering, progress in the molecular sciences is providing an ever-stronger foundation for systematic molecular systems engineering. In the last dozen years, the means for designing and building molecular components have become powerful indeed.

The component technologies are already in hand, and it's time to learn how to put them together. And this is a task that calls for organization.

CHAPTER 14

How to Accelerate Progress

EARLIER I OFFERED THE TWIN parables of the elephantologists and automobilists, two research communities that shared a curiosity-driven style of organization. The elephantologists produced a coherent result— an increasingly detailed understanding of elephants, with descriptions templated on the elephants themselves. The automobilists, by contrast, produced nothing that resembled an automobile, because curiosity-driven inquiry doesn't lead to system-level designs, and without a coherent design concept there isn't a template to guide research, no basis for producing coherent results. The automobilists developed components (wheels, gears, engine-like things, and the like) but these were seldom suited to automotive applications and didn't fit together to form a functioning system. Research on brakes never became fashionable, and no one noticed a need for differential gears.

The moral is that no matter how research-intensive a project may be, work coordinated around concrete engineering objectives will eventually be required to produce concrete engineering results. Whether the leaders call themselves engineers doesn't matter (and leadership may at first be diffuse); what matters is that the leadership community knows how to coordinate multiple groups to produce, first a comprehensive set of compatible technologies, and then working systems.

This is true in high-energy physics, with its billion-dollar–scale accelerators, detectors, and data processing systems. This is true in space science, with its billion-dollar–scale spacecraft and Mars rovers. This is likewise true of the ten-million-dollar–scale research that yields advanced DNA sequencing systems. All these systems are products of science-intensive engineering in the service of science.

Today, AP molecular systems engineering needs this kind of focused effort along with the precursor developments necessary to make it effective. The time is ripe to take this next step. The world research community has advanced far along the road, yet there's work to be done in developing a more comprehensive set of compatible techniques and components.

This kind of technology platform—a basis for more effective and systematic engineering—can serve many purposes in both science and practical applications. When put in perspective, of course, the most important application will be early-generation systems for guiding next-generation molecular assembly, supplementing self-assembly with soft stereotactic control.

ROADMAPPING FOR PATHFINDING

The semiconductor industry provides a model for coordinating research to advance the technology of an entire field. What's more, the achievements of semiconductor engineers give us a sense of the potential scale of results, for it was their work that brought us nanoscale digital information systems and today's Information Revolution.

Roadmapping has been a key to their success, a way to coordinate research and development, and, in particular, a way for people to formulate complementary plans and share a confidence that their successes in developing parts can bring the rewards that can only come from contributing to a working whole. Roadmapping helps to solve this problem of plan coordination.

Roadmapping Moore's Law

For decades now a formalized roadmapping process with a fifteen-year horizon has helped to drive the Moore's Law revolution in electronics. This cooperative, industry-spanning effort produces and updates the massive ITRS document, the *International Technology Roadmap for Semiconductors* (ITRS).

This roadmapping process has sped progress by coordinating efforts and reducing the risk in doing what's necessary.

Each generation of lithography has required a set of improved and compatible technologies for light sources, optics, masks, steppers, resists, and a raft of other process technologies. To invest with confidence in the development of a shorter-wavelength light source (of adequate collimation and brightness), and at a particular time, one must have confidence that there will be a market—and a market will exist only if other companies deliver the compatible optics, masks, steppers, and all the rest, and only if major producers plan to build fabrication facilities based on this suite of technologies.

The ITRS process helps provide that confidence and crystallizes the shared expectations into formal, public documents. It also helps to develop mutual expectations with names, faces, and corporations attached.

Roadmapping Quantum Information Systems

The Quantum Information Science and Technology Roadmap (QISTR) shows a way to do roadmapping in a domain where the nature of practical implementation technologies remains uncertain. Participants in the ITRS can safely assume that silicon will rule for years to come, but the QISTR collaboration faced a range of fundamentally different competing approaches: quantum bits represented by the states of (pick one or more) trapped atoms in a vacuum, spin states of atoms embedded in silicon, nuclear spins in solution-phase molecules, or photons in purely photonic systems. These approaches differ radically in scalability and manufacturability as well as in the range of functions that each can implement.

The QISTR document must rise to a higher level of abstraction than ITRS. Rather than focusing on performance metrics, it adopts the "DiVincenzo promise criteria" (including scalability, gate universality, decoherence times, and suitable means for input and output) and through these criteria for essential functional capabilities, QISTR then compares diverse approaches and their potential to serve as more than dead-end demos.

QISTR shows how a community can explore fields that are rich in alternatives, identifying the technologies that have a genuine potential to serve a role in a functional system, setting others aside as unpromising. (For example, solution-phase NMR techniques have been used to demonstrate principles, but they can't be scaled to more than a few bits, and so as a basis for further development they don't make the cut.)

Roadmapping Paths to APM

The development of atomically precise nanotechnologies is an ongoing process and, I've mentioned, the first roadmapping effort directed toward advanced, system-level objectives (the Battelle-led project, hosted by several US National Laboratories). The resulting document, *Productive Nanosystems: A Technology Roadmap* (PNTR), surveys paths that lead from current atomically precise fabrication technologies toward atomically precise manufacturing. From the Executive Summary:

> This initial roadmap explores a small part of a vast territory, yet even this limited exploration reveals rich and fertile lands. Deeper integration of knowledge already held in journals, databases, and human minds can produce a better map, and doing so should be a high priority. Some research paths lead toward ordinary applications, but other paths lead toward strategic objectives that are broadly enabling, objectives that can open many paths and create new fields. These paths are the focus of this roadmap. They demand further exploration.

PNTR occupies a middle ground: Like QISTR, it explores alternative sets of technologies, yet unlike QISTR, the general nature of workable systems is clear. To this extent, AP technologies resemble semiconductor technologies, yet ITRS, unlike PNTR, draws on long experience in which each generation of technologies has provided a template for the next. Roadmaps evolve with their technologies.

Science-Intensive Engineering

The optimal mode and balance of organization and management depends on the nature of objectives and the nature of the field. In semiconductor technology there's a funnel that opens toward basic science that studies novel phenomena in new materials and processes, and narrows toward focused problem-solving for systems that will play roles in next-generation chips, with well-defined requirements for lithographic resolution, gate-oxide properties, and all the rest. The broad end of the funnel, basic research, is only loosely polarized toward the narrower end, where research becomes development, and incremental production process improvement.

When a technology is far from maturity—far from producing successive generations of products—there can be no good understanding of its requirements. At the inception of a technology, there is no well-structured funnel like that of semiconductor technology—the basic science may be advanced, and yet at the broad, basic-research end of a nascent development funnel, the end goals may not yet be clear. Even at the earliest stages, however, when almost all of the work will be (and should be) science-intensive and largely curiosity driven, that work can be loosely polarized, oriented in a direction, and linked in some way to criteria and metrics for prospective engineering applications.

Molecular systems engineering today is at that early stage, but deficient in a shared sense of orientation. Advances in general knowledge can help advance progress, but the pace of progress today will be set primarily by advances in understanding potential objectives and advances in knowledge of how to achieve these objectives. Molecular systems engineering is emerging. The question is how to proceed, in what directions, and with what level of ambition in delivering results.

The Road Ahead in Molecular Systems Engineering

The case at hand is unusual for the reasons we've seen. Open paths lead toward an enormous reward, toward classes of physical systems beyond the usual horizons of experimental research, and yet for a range of technical reasons objectives that are further from reach can be more easily explored and analyzed than objectives that are closer at hand. For example, machine components based on rigid covalent structures cannot yet be implemented, yet are already easy to design and model using standard computational chemistry software. This unprecedented situation calls for applying exploratory engineering methods across a spectrum of technological levels in order to develop a better understanding of potential pathways and effective lines of research.

As noted above, some of the areas of greatest promise center on macromolecular structures, self-assembly, and computational methods for design and modeling in support of this work. Other areas of research complement this line of development and they embrace advances scattered across the broad fields of nanotechnology as presently defined—some are atomically precise, but others are not.

We've also seen that a continuum of technologies stretches from pure self-assembly, to stereotactic methods that apply both self-assembly and loosely controlled positioning, all the way to tightly controlled motions of the sort required for advanced APM. This gradient of fabrication methods and products can be crossed through a series of modest and rewarding steps. The road isn't short, but reaches the full distance.

Where development is concerned, the great challenge today isn't technical. What's most needed today is something human: a shift in the conversation, a shared sense of opportunity, and a research environment that supports a more informed and coordinated focus on engineering objectives.

———

APPENDIX II takes a closer look at the technology gradient that reaches from present laboratory capabilities to APM-level technologies. It outlines how several dimensions of progress naturally fit together, including advances in materials, devices, systems, and means for design. The resulting picture shows how long strides can be taken on a path with closely spaced stepping stones.

Current AP Capabilities, Pathways, and APM-level Technologies

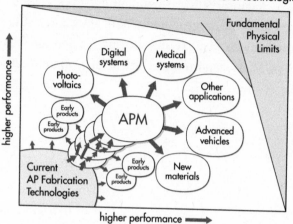

FIGURE 3: Pathways to APM-level Production and Products

This discussion has several related take-away messages that rise above the specifics. One is that many paths lead forward, are broad, and are far from fully explored. As happened a few years ago in structural DNA nanotechnology, new insights may abruptly and unexpectedly expand capabilities. Another is that the pace of overall progress will be set by the pace of progress in techniques (physical and computational) that enable faster, more predictable design cycles. A third is that as the design cycle shortens, the pace of progress could well accelerate to a degree that will take much of the world by surprise, even if research programs are highly transparent.

In light of this prospect, it isn't too early to consider potential opportunities, including implications of advanced APM for addressing global problems unfolding today. These implications include solutions to problems as intractable as climate change. Nor is it too early to consider potential problems, the disruptive implications of APM technologies for society, economics, and international affairs. These implications are profound, and as challenges draw near it will be essential to understand the basic choices in order to avoid needless risks and manage a potentially catastrophic success.

From where we stand today, it would be unwise to rely on APM for quick solutions to anything, but it would equally unwise to rely on prolonged delays. A conservative assessment must respect the full range of uncertainties, and this range includes both slow and swift progress.

Let's return now to what physics and engineering can tell us about the potential of APM, as seen through the lens of human needs and desires.

PART 6

BENDING THE ARC
OF THE FUTURE

CHAPTER 15

Transforming the Material Basis of Civilization

WHAT ARE THE NATURAL practical applications of the physical capabilities of APM-level technologies, the applications that will matter to people and the Earth? Exploratory engineering methods enable only a partial survey, but what can be seen is enough to call for a radical reassessment of prospects for the twenty-first century. In practice, a radical reassessment must begin incrementally, first encouraging inquiry into key questions (regarding potential technologies, timelines, implications, and policy options), and then hedging bets (intellectual, technological, financial, political, and so on) in response to what the answers suggest.

As outlined in Chapter 11 the key physical capabilities include low-cost production of higher-performance materials, leading to higher-performance components and products of essentially every kind: stronger, lighter structures, more efficient engines, greater safety, lower emissions, and vastly greater computational power. The advantages of

atomic precision spill over into medicine, too, where molecular interactions are crucial.

Cost is a crucial concern for technologies, and unless costs are affordable even the highest-performance technologies will languish as
laboratory curiosities. In the case of APM-level production technologies, however, high performance is virtually synonymous with low-cost
production. Technologies that offer higher than high-end performance
coupled to lower than low-end costs have been a rare and disruptive
market phenomenon, yet this is what APM promises to deliver, and not
just in one area of application.

These general characteristics of performance and cost—to say nothing
of wholly new capabilities—will drive far-reaching changes. The balance
of this chapter outlines some of the natural impacts in several important
areas of application. In the spirit of exploratory engineering, these descriptions aim to present, not the best possibilities, in a technological
sense, but rather a low-end range of applications of intrinsically high-
performance technologies, in order to describe a level of performance
that can be expected with confidence.

A FEW WORDS ABOUT CONSUMER PRODUCTS

APM-level technologies will expand the range of accessible products
that can be driven, worn, or used in sports or in daily life at home—
conveniences, entertainment systems, and a broad range of consumer
products. The reader will readily see some of the impacts of new production technologies in the consumer sphere; these start with deep cost
reductions, but also embrace a range of both higher-performance and
entirely novel consumer technologies.

The nature and impacts of new consumer technologies are often unexpected. The impact of automobiles on the structure of cities and daily
life, for example, hadn't been predicted, and no one predicted the nature
of the modern Internet or what it has become as part of daily human experience. Speculations in this area may be worth undertaking if the results are regarded with sufficient skepticism, but here, such speculations
would be out of scope.

Consumer goods pervade our experience of life, yet they aren't part of its deeper physical basis. My concern here is with developments at that deeper level, where some of the most basic driving forces for change can be expected with substantial confidence.

TRANSFORMING THE MEANS OF PRODUCTION

From where we stand today the coming transformation can best be understood through contrasts with current industrial technologies. As we've seen, the primary contrasts emerge from just two basic characteristics of APM-level technologies: the nanoscale size of components and the atomic precision of processes and products. As we've seen, there are several key implications of these from an applications perspective.

The first, nanoscale size enables extreme productivity as a consequence of mechanical scaling laws. In addition, small-scale, versatile, highly productive machinery can collapse globe-spanning industrial supply chains to just a few links, from raw materials to refined feedstock materials, from feedstocks to standardized microblocks, and then from microblocks to products that play roles as different as solar cells, spacecraft, car engines, concrete, computers, and medical instruments. Short supply chains and flexible production can enable radical decentralization.

The second, atomic precision starts with small-molecule feedstocks, atomically precise by nature and often available at a low cost per kilogram. A sequence of atomically precise processing steps then enables precise control of the structure of materials and components, yielding products with performance improved by factors that can range from ten to over one million. And because precise processing embraces both products and by-products, APM-based systems need not produce hazardous wastes.

Both industry and APM produce physical products, yet their contrasts point to radical differences. As we've seen, the Information Revolution provides an alternative model.

Producing patterns of atoms using APM-based technologies once again resembles producing patterns of bits using information technologies. Rapid production based on multipurpose, scalable platforms;

independence from long, specialized supply chains; the potential for radical decentralization; the pivotal role of software and online data; new products without costly new physical capital; low marginal costs of production and distribution; the potential for rapid, global deployment of new products—all these characteristics are shared by both APM and information technologies, yet all contrast sharply with the characteristics of modern industry.

TRANSFORMING INFORMATION TECHNOLOGIES

APM-based production will boost the Information Revolution itself. At the physical level of information technologies, where rapid change is already routine, APM-based production can boost ongoing trends, carrying them further (and perhaps faster) than expected.

Modern computation, telecommunications, sensing, display—and all that they imply for daily life, economics, and the structure of society—have grown out of the ongoing, exponential, Moore's Law revolution in manufacturing that now enables single chips to outperform what had been called supercomputers just a decade ago.

This exponential progress cannot continue forever, and for today's semiconductor device manufacturing methods, the limits are already in sight. Since 1970, transistors have shrunk from dimensions of ten thousand to just ten atomic diameters, yet from the beginning they've been built by methods that inherently lack atomic precision. APM-level technology can go much further, but the size of the step will depend on the performance of other technologies at the time, and those technologies will be a moving target. Indeed, specialized AP fabrication processes will likely enable hybrid-technology chips, smoothing the transition.

Estimating potential advances measured against today's technologies is more straightforward. One can expect reductions in energy consumption (comparing low-power processors with comparable capacity) from milliwatts to nanowatts, and reductions in processor scale from millimeters to micrometers—in volume, a factor of roughly one billion. Increases in single-core processor speeds will depend on the speed of digital devices, and fundamental physical constraints suggest that the remaining potential increase in speed will be less than the factor we've already seen in the evolution of transistor-based machines.

Telecommunications costs have fallen even faster than the cost of computation. Current technologies enable transmission of many terabits per second through an optical fiber just tens of microns in diameter. While fiber data capacities continue to climb, free-space transmission is burgeoning; at the present (2012), WiFi delivers tens of megabits per second and industry plans to multiply this by a factor of one hundred; the physical limits have yet to be reached.

Here, too, APM-level technologies can both increase capacity and lower costs. On a per-unit-mass basis, the critical electronic guts of current telecommunication systems cost more than $1,000 per kilogram, leaving room for thousand-fold cost reductions. Potential performance improvements are harder to estimate, but will likely be large.

Digital information systems interact with the world through sensors, displays, and control signals. Regarding sensors, laboratory devices and high-end commercial systems set a lower bound on what to expect; in cameras, for example, device sensitivity can already approach quantum-limited, photon-counting performance, while a range of chemical sensors can detect single molecules (expect fast, nearly zero-cost DNA readers). Regarding image displays, arrays of devices for emitting light and changing reflectance already reach the resolution limits of human perception; the scope for improvement includes wearable devices and three-dimensional, window-like image quality. Regarding control of mechanical devices, the most striking advances won't be in the controllers, but instead in the range of devices themselves.

Putting the pieces together, potential applications of APM-enabled information technologies include the realization of extreme forms of what has already been imagined, including ubiquitous computing, networking, information services, and surveillance. How much of that potential will be realized remains to be seen—or better yet, and thinking less passively, remains to be considered, discussed, proposed, negotiated, legislated, and implemented.

TRANSFORMING INFRASTRUCTURE

Industrial Equipment

A natural, cross-cutting impact of APM-based production will be reductions in the purchase and operating costs of industrial-level capital goods, including the machinery used to move things, build things, and provide utilities such as water and electricity. Later sections discuss utilities, but as enablers of the rest, construction and transportation come first.

Construction

In construction, APM-level technologies will improve the performance of materials, structures, and functional components while reducing the costs of their production and use. Because most structural materials used in construction already have low costs per kilogram (concrete, for example), the impact of cost reductions may be modest and will be relatively sensitive to eventual APM costs. Functional materials, by contrast, have great scope for improvements.

For example, vacuum aerogels, though costly and fragile today, can equal common glass fiber insulation at only one-tenth the thickness. With advanced fabrication, insulating materials as good or better can be both robust and inexpensive. Services such as heating and cooling have less improvement (both can already approach thermodynamic limits), but once again, products better than today's best can be produced at sharply reduced costs and hence more widely deployed (heat pumps and air-to-air heat exchangers save energy, but current costs limit their use).

Costs of assembly represent much of the cost of construction. Here one can see potential for substantial improvements through the production of low-cost, prefabricated, yet precisely customized segments of larger structures—lightweight, easy to move, and designed for easy assembly.

Transportation

Lower cost production, stronger, lighter materials, engines with higher power density and efficiency, zero-emission energy sources—all these can lower the cost of transportation, including its environmental impacts.

The greatest advantages will appear where costs are high and performance is critical, in aerospace systems in general, and space systems in particular. The cost of access to space today has surprisingly little to do with energy requirements and has everything to do with the cost, mass, and reliability of vehicles. Decades ago, the cost of spaceflight blocked the dream of space settlement, but that barrier will drop.

TRANSFORMING ENERGY, RESOURCES, AND AGRICULTURE

Industrial equipment, construction, and transportation constitute a large portion of modern economies, and changes in their material basis will have pervasive effects, including changes in resource supply and demand.

Energy

On the demand side, APM-level technologies will increase energy efficiency across a wide range of applications and sometimes by large factors. Improvements in power-conversion efficiency, vehicle mass, thermal insulation, and lighting efficiency are examples. In ground and air transportation, the accessible improvements include ten-fold reductions in vehicle mass and a doubling of typical engine efficiencies. Taken together, improvements like these enable deep demand reduction, while lower costs of production can enable faster replacement and upgrade of systems already in place. Other attractions (cleaner, safer, higher performance, and so on) would likewise spur replacement of existing capital stock.

On the supply side, improvements in costs and technologies can enable extensive and potentially rapid replacement and upgrade of energy infrastructure. The energy industry is highly heterogeneous, but every sector is capital intensive; reductions in the costs of physical capital will lower the cost of new installations of all kinds, facilitating replacement of capital stock at rates that could surpass any in historical experience.

In particular, improvements in costs and technologies will boost solar electric power while making coal-fired power plants (2,300 today) vulnerable to fast replacement. Indeed, when combined with efficient, inexpensive APM-enabled technologies for interconverting electrical and chemical energy, solar energy can provide both baseload electric power and liquid fuels on a global scale; as outlined in Chapter 11, Earth-abundant elements can be used to make efficient, nanostructured, thin-film photovoltaic cells and the resulting electrical energy can be efficiently stored in conventional liquid fuels and recovered as electric power for use in vehicles or fixed energy infrastructure.

Turning to questions of capacity and impact, to meet current global energy demand (about fifteen terawatts, including wood and dung burned for heat) would require about 0.2 percent of the Earth's land area, or about 1 percent of the area now used for grazing and crops. With sheets of tough, abrasion-resistant composite materials used in place of fragile photovoltaic panels, rooftops and roads could provide much of the area required.

Raw Materials

Raw materials and their uses are diverse today and will be diverse tomorrow, and this complicates the question of how APM-level technologies are likely to affect raw materials demand. Chapter 11 explored this relationship and concluded that APM can reduce the demand for scarce resources in two ways:

By enabling less massive products to perform common functions (architectural and mechanical structures, electrical wiring, electronic systems, and so on).

By enabling abundant elements (primarily hydrogen, carbon, nitrogen, oxygen, aluminum, and silicon) to substitute for scarcer materials (copper, nickel, cobalt, zinc, tin, and others) in most applications, and with better performance.

These changes can greatly relieve the pressures of resource scarcity, now a growing cause for international tension. For materials that are still in demand, improvements in the cost and performance of industrial equipment can reduce the costs of mining, refining, pollution control, and remediation. (Note that earlier estimates of APM production costs took a narrow view of the economic context, and hence did not take account of prospective reductions in the costs of required raw materials and energy.)

Water

The growing scarcity of water for human and agriculture use ranks high on the list of global problems, a problem made worse by population growth, environmental degradation, and climate change. Abundant energy and improved, lower-cost capital goods can address this problem directly by lowering the cost of desalination and water transportation, drawing fresh water from the sea. Atomically precise fabrication can produce membranes with tailored water-transporting molecular pores, enabling higher performance systems for reverse-osmosis desalination, and (of critical practical importance) can lower the cost of producing, cleaning, and recycling membranes and other filters and surfaces subject to fouling.

Agriculture

Agriculture consumes over 80 percent of the world's fresh water supplies, and also pollutes them, tying resource and environmental concerns to the potential for improving agricultural methods.

Across most of the twentieth century, grain production grew faster than population, but that trend has flattened since 1990 and begun to decline, while food prices have recently trended sharply upward. Concerns about the impact of climate change and water shortages have reinforced fears of a food shortage. Once again, APM-based production capacity improves the prospects for meeting demand while respecting environmental constraints.

The world has increased food production by three main methods: by expanding the area under cultivation, by applying chemical fertilizers, and by planting crops with higher yield. Each of these methods faces diminishing returns as new fertile lands become scarce, the incremental benefits of fertilizers decline, and the potential for plants to be more productive approaches limits determined by temperature, soil, and available water—a limit that can destroy crops entirely in drought years.

Enclosed agriculture (greenhouses, for example) can greatly increase and stabilize yields, largely freeing agriculture from constraints of temperature, soil, and water. Compared to unprotected environments, where the vagaries of location and climate determine growing conditions, using controlled environments can commonly raise the productivity of land by a factor of ten or more.

Optimizing growth conditions requires enclosures that control temperature (usually warm, never too hot), humidity (usually high, but not saturated), and sunlight (typically bright, but diffused, not direct), and that provide soil with ample nitrogen, phosphorous, and potassium. A well-controlled enclosure can also exclude pests without using pesticides, and can recycle nitrogen and phosphorous, retaining them to fertilize crops without contaminating streams.

To accomplish this on a large scale requires an abundance of physical capital: the structural components for building the enclosures, the equipment they must contain—pumps, pipes, and filters for water reprocessing, heat pumps and thermal storage to regulate temperatures, and finally, sources of power to make them all work.

The rewards of expanding the use of enclosed agriculture would include higher yield per hectare, but also better food quality, freedom from pesticides, extended growing seasons (in many regions, year-round production), freedom from constraints of soil quality and available water, and protection from drought. From a biospheric perspective, benefits would include reduced water demand and contamination, and a way of supplying human needs for food while reducing the overall footprint of agriculture and relieving pressures that drive the deforestation of Amazonia. Cleanly increasing agricultural yields by a factor of ten would change human life and the face of the Earth.

TRANSFORMING ENVIRONMENTAL CONCERNS

Transforming the material basis of civilization can transform the impact of human beings on the Earth, perhaps for the better. Although lowering costs could enable greater destruction, I am persuaded that cleaner, low-impact technologies can lead to better net outcomes, provided that people who care about the fate of the Earth keep pushing to make it better. The prospect of greatly expanding production while simultaneously reducing environmental impact offers an opportunity to resolve some of today's most intractable conflicts and to set a new pattern for how human civilization coexists with the rest of our world.

We've already seen some of the ways in which APM-based production can diminish our environmental impact, for reasons that include more frugal use of materials and the ease of avoiding outputs that pollute the air, water, or soil. To this add an energy infrastructure based on solar energy with collectors deployed in locations and configurations chosen not to reduce an already low cost, but instead to reduce impacts on ecosystems.

Radical abundance can serve many purposes, including some that had seemed incompatible. Rather than thinking of radical abundance as "more," or as a shift in a familiar trade-off, it's better to think in terms of shifting entire trade-off curves upward, enabling outcomes that are better from a range of perspectives, including perspectives that seemingly clash. This can (and should, and might) change politics.

Environmental Restoration

Physical challenges can make environmental restoration costly or ineffective. To do a good job of repairing a strip-mining scar, for example, it may be necessary to move many millions of tons of rock and soil. Efforts to remove toxic chemicals and heavy metals from soils today are both expensive and incomplete. All points on the trade-off curves are unsatisfactory.

Once again, APM-based production can improve the cost and performance of the necessary equipment, and beyond this, APM-level technologies can provide new capabilities for capturing and sequestering toxic materials from groundwater and soil, and for more subtle challenges of remediation.

Reversing the Primary Driver of Climate Change

With APM-based production and products, energy sources and most energy uses can be engineered for a zero net carbon footprint; liquid hydrocarbon fuels, for example, can be produced using hydrogen from water and carbon from recycled CO_2.

Reversing the effects of past emissions, however, would require atmospheric CO_2 capture on an enormous scale—some three trillion tons—a remediation task that appears to be beyond the capacity of the industrial civilization that created the problem. The energy requirements are daunting. Even if CO_2 is captured with high efficiency, the necessary energy is roughly equivalent to ten years of today's total global electric power production.*

From an APM perspective, however, this challenge seems manageable. APM-level technologies can provide thermodynamically efficient means of capturing and compressing CO_2 from the air, while the required energy could be provided in ten years by scattered photovoltaic arrays with a total area 0.5 percent as large as the Sahara desert (the equivalent of a single 200×200 km array). By these means, APM-based production could provide sufficient carbon capture capacity to return the Earth's atmosphere to its pre-industrial composition in a decade, and at an affordable cost. This places a solution to the CO_2 problem within reach—but only eventually, after an unknown and risky delay.

———

TAKEN TOGETHER, unexpected prospects for averting the collision between civilization and the limits of the Earth offer reasons for hope where hope has been scarce.

———

* This reflects the irreducible thermodynamic work required to compress the entire industrial-era CO_2 surplus from a dilute gas to liquid densities (roughly 10^{21} Joules).

TRANSFORMING SECURITY TECHNOLOGIES

Security applications of APM-based technologies present a more ambivalent prospect. Advances in technology have always found their way into weapons, and replacing one generation of weapons with the next has had mixed effects on human life.

"Security technologies" include means of offense, defense, and surveillance applied to purposes as different as war, neighborhood protection, and political repression. Looking toward future security technologies, potential applications range from unconstrained, deadly, and unaccountable force, to gathering information subject to well-crafted rules for protecting yet not threatening civil society. Today, security technologies involve both physical and information systems. Here I'll discuss only the physical aspects, postponing questions at the level of policy and strategy to a later chapter.

Platforms and Surveillance

As noted above, APM-level production can amplify trends that are already in progress, expanding the potential for ubiquitous, networked, computing-intensive systems of sensors that would extend an already growing surveillance infrastructure. Nothing new is needed for fixed, ground-based surveillance networks, and potential improvements in the cost and performance of mechanical and aerospace systems highlight the potential for rapidly deployable drone networks on land, sea, and air.

Small-scale components and systems open new possibilities. For perspective, consider that a one-gram platform built with advanced technologies could provide teraflops of computational power (and much more, in bursts), together with a million-terabyte data storage capacity and better-than-human sensors, all with a power demand comparable to that of a cell phone on standby. Now consider what could be done if devices of this class cost about $1.00 per kilogram and could be delivered by small drone aircraft. $100 billion would buy roughly one device of this sort per square meter of land area, worldwide.

Application of Force

As the history of military drones has shown, surveillance systems can enable lethal applications. In a world with low-cost production and potentially intensive surveillance, however, lethal weapons may become less important.

There's been a trend toward reducing lethality. High-yield thermonuclear weapons, for example, achieved the apogee of deaths-per-dollar in the 1960s, and in recent decades precision munitions have become more popular than carpet bombing. Applying less force with greater precision has reduced lethality, though seldom entirely avoided it.

The reasons for using lethal weapons have been partly economic. Per unit of cost, non-lethal weapons are typically less effective—as a deterrent, the threat of a bullet can be more effective than a cloud of tear gas, and bullets are cheaper. However, with inexpensive surveillance and weapons production, large-scale, highly effective deployments of non-lethal weapons will become more affordable, and a natural direction for advances in non-lethal weaponry would be to apply coercive force with greater control while avoiding physical harm.

Force has always been the foundation of power, and radical changes in the means of applying force should raise deep concerns. In particular, the deployment of increasingly effective non-lethal weapons tends to relax moral and political constraints on coercion. In a domestic context, freer use of coercion could open the door to abuse on a scale that subverts society. In a military context, these same capabilities could lower the threshold for actions that risk broader conflict and war.

What heightens APM-related security concerns and makes them more complex is the context in which they will emerge, a context that promises turbulent change. I will return to this topic in the following chapter.

TRANSFORMING MEDICINE

APM-level technologies are a natural fit to applications in medicine because medicine is rooted in biology and biology is based on atomically precise structures. Biomedical research today relies heavily on atomically precise devices, both for knowledge and interventions.

Knowledge and Diagnosis

As is often true in science, advances in technology set the pace of progress in biology, but in biology much of the technology is based on adapting biomolecular devices. Genome research relies on DNA sequencing methods that use AP mechanisms borrowed from cells, and the same is true of transcriptome, proteome, and metabolome research—which is to say, comprehensive studies of RNA molecules transcribed from the genome, the proteins those molecules encode, and the metabolic products of those protein's functions. These studies reveal patterns that vary from tissue to tissue, from youth to age, revealing changes in states of health and disease in unprecedented detail.

Today, the costs of substantial comparative studies of this kind can mount into the million-dollar range. Advanced AP instrumentation, packaged in hand-held devices, could gather more complete data more quickly and at a negligible cost per use. At that point, formerly costly scientific techniques become routine clinical methods for monitoring and diagnosis. These examples illustrate the natural fit between AP technologies and biomedical research and give a taste of the potential.

Fighting Infections and Cancers

When the body clears infections and cancer cells, molecular recognition—a specific fit between molecules—directs its attacks. Both the innate and adaptive immune systems play roles, but they are not always able to mount a response quickly enough or thoroughly enough or precisely enough, and in some instances they simply fail to respond.

Molecular-level studies can identify the distinguishing features of pathogens, and advances in AP technologies could speed this process and enable a rapid response. In one approach, the next step would be to manufacture corresponding antibodies—or specialized analogues—for treatments that parallel the use of gamma globulin to temporarily boost a patient's immunity to hepatitis A or measles. The ability to speed development of agents like these and to expand large-scale production quickly could avert the threat of emerging pathogens.

Cancer cell lines differ only subtly from normal cells, making immune recognition more difficult and ensuring that chemotherapies will have systemic toxicity. Research in nanomedicine has made progress in targeting nanoscale particles to deliver chemotherapeutic agents more selectively; more advanced AP technologies could produce devices that test several cell characteristics before triggering the delivery of a cell-killing payload; with refinement, devices based on this principle would enable nearly perfect targeting of cancer cells, avoiding significant side-effects.

Supporting Correction, Repair, and Regeneration

Experimental biomedicine (usually in experiments performed with mice) has increasingly turned to interventions that heal, repairing and regenerating tissues rather than merely destroying invaders. Altering patterns of gene activation—suppressing some outputs, increasing others—can have profound effects on cellular metabolism, growth, and differentiation with results that include converting scar-tissue cells into cardiac muscle.

Note that none of the above developments requires controversial gene therapies or cells salvaged from human blastocysts. Increasingly advanced medicine, though, will doubtless generate increasing controversy, sometimes regarding its means, but often regarding its ends.

Technologies at or approaching the APM level will increase the reach of medicine in innumerable ways, with consequences in areas that range from medical ethics, practice, and costs to demography and long-term prospects for national budgets.

A BANQUET OF INDIGESTIBLE TRUTHS

In this chapter, I've outlined a range of technological prospects with uncomfortably broad implications. Viewed from the standpoint of physics and engineering, their basis is straightforward, resting on well-established knowledge regarding molecular interactions and mechanical scaling laws, and from this standpoint, many implications of small-scale machinery and precise control of molecular motions are almost obvious—high productivity, better materials, and better and lower cost products follow directly—and through conservative exploratory engineering studies, a wide range of more specific conclusions follow.

The conceptual problems arise when we trace purely technological chains of implication forward far enough to reach matters of direct human concern. At that point, when viewed from a human perspective, facts about the physical potential of atomically precise fabrication technologies erupt into a mass of potential consequences, presenting an intellectual meal that I, for one, find too large to fully digest.

———

THE PRESENT CHAPTER built a bridge from engineering-level implications to applications that matter to people and nations, examining these piece by piece, and not as a whole. In the next chapter we will further widen our view and attempt to assess how new opportunities and risks could combine to reshape economic and societal challenges on a national and international scale.

The result will offer few answers, but may help to sharpen the questions.

CHAPTER 16

Managing a
Catastrophic Success

HOW WILL THE APM revolution change the world? There can be no detailed or confident answer today, yet some of the potential driving forces are visible, and with these, the tentative outlines of a range of potential directions. Even with only this limited knowledge, we can be confident of some aspects of the answers, considerations that can guide conversation toward realism.

Realism begins with physical law and the timeless landscape of technological potential, and within this framework, an understanding of some of the capabilities of APM-level technologies. Exploring this topic (and the nature of the exploration process itself) has been the subject of much of this book.

Realism also includes uncertainties, a view of the world that takes account of unknowns. Some uncertainties are, in a sense, reliable, because one can rely on having no confident answers. This is a familiar aspect of planning; attempts to quantify "market risk," for example, reflect irreducible uncertainty that must be considered in rational planning, and similar concerns arise in every sphere of human action that looks toward the future.

Regarding technologies, unknowns include how and when APM will emerge, and the nature of applications that have not yet been envisioned. Beyond these are uncertainties regarding responses and outcomes, how (and to what extent) APM's potential capabilities will be applied, and by whom, for what purposes, and with what effects.

Finally, realism requires efforts to develop *coherent* views of the future—not predictive or even comprehensive scenarios, which can't be had, but scenarios that aren't inherently and obviously inconsistent. For instance, it would be wrong to imagine just a bit of APM sprinkled into an otherwise conventional world, because APM-level production technologies will deliver diverse, compelling capabilities in what amounts to a single package. For example, in considering a future in which petroleum has lost its value, it would make no sense to assume that APM-enabled powers will still compete for access to oil fields. Likewise, in considering a future where the technologies of radical abundance disrupt employment, it would make no sense to tacitly assume this will happen in a world locked into our current levels of scarcity. Incoherent scenarios would engender incoherent policies, leading to needless risks and missed opportunities.

My aim here is to encourage inquiry, thought, and conversations that lead to a more realistic, coherent view of our future, and through this, in time, to choices that may tip the balance toward favorable outcomes. To that end I'd like to offer some observations on how prospective capabilities could enable solutions to global problems, with attention to the challenge of closely linked disruptive effects.

In the following chapter I'll turn to questions of security and national interests that will arise not only with the arrival of an APM-level technology base, but also before that, from its anticipation. Regarding potential strategic responses, it seems there's a fork in the road.

QUESTIONS OF PACE AND DIRECTION

How Far? And How Fast?

Physics and engineering can directly address questions of APM's physical potential, and this potential includes implications for the potential pace of deployment. Thus, exploring answers to the engineering question, *How far?* can inform judgments of *How fast?*

When?

This question differs from the question, *How fast?* because the answer depends on development pathways (alternatives, difficulties, shortcuts) and the scale and focus of research (among other considerations). The pace of past progress in AP fabrication can give no more than hints regarding the likely pace of progress going forward.

With What Result?

This asks an open-ended question that cannot have a specific, confident answer (*"This* will happen"), yet the range of possible answers can be constrained by technology-centered facts, not only because physics imposes limits, but also because in a competitive world, opportunities sometimes force a response.

Many Feet on Many Accelerators

Regarding the pace of technology development in world society, the diversity and independence of potential actors constrains likely outcomes by multiplying potential accelerators while providing no single location for brakes. Regarding the technologies themselves, the diversity and independence of potential paths forward have similar implications, multiplying potential shortcuts and routes around any particular roadblock.

It's easy to imagine strong programs emerging, but consider for a moment the scope of potential bottom-up changes in science and the low threshold for shifts in direction.

Today, in AP fabrication research many groups in many countries are pursuing many paths for reasons that include scientific curiosity and practical applications in materials science, chemistry, biology, and medicine, and this research has brought ongoing rewards. An approach more focused on developing AP tools and systems, however, promises a more quickly expanding range of rewards in these same areas, and there is no barrier to going further. At the entry level, all that's required is a shift in focus by groups working within their fields of expertise—continuing to work, for example, with a familiar class of self-assembling molecular structures while choosing specific research targets with system-level objectives in mind.

Where might one expect to see research programs with a stronger focus and a more coordinated approach? Candidates include nations that have funded programs in generic, non-AP nanotechnologies, which were often established in hopes of progress toward APM-level results. By this criterion, candidates include the United States, China, Europe (both in individual countries and in Europe as a whole), Russia, Japan, India, Korea, Australia, Singapore, Malaysia, Thailand, and Brazil (this is a partial, illustrative list).

Both near-term and long-term objectives point in the same direction. A shift in research focus toward molecular systems engineering (hence APM pathway technologies) would serve near-term objectives in the range addressed by research today, but with the added lure of growing and, ultimately, immense long-term rewards. Such longer-term rewards, once understood, are compelling. As we've seen, the rewards of APM-level technologies not only include areas like medicine, materials science, and energy, but also extend to solving problems at the level of global economic development and climate change. Conversely, viewed through the lens of national competitiveness, the cost of missing out on the APM revolution could be disastrous.

In making a net assessment, both opportunities and competitive pressures must be weighed against the potential costs and risks of disruptive change. Some of these costs and risks are avoidable, some seemingly not. In considering benefits, costs, and risks, however, we face a fundamental asymmetry between accelerating progress and attempting to slow it. Adding an increment of effort pushes global progress forward, but withholding an increment doesn't push progress back. Instead, it amounts to stepping back toward becoming a spectator, with less opportunity to join in efforts to realize potential benefits and mitigate problems. From a competitive perspective, stepping back increases risks that stem from falling behind while also reducing potential influence on cooperative measures that address collective, systemic, international risks.

In light of these considerations, what can one say about the likely time of emergence of APM-level technologies? Although potential development pathways are clear in outline, specific dates and milestones are beyond prediction; this is why the discussion here centers on driving forces, patterns of action, and relative speed while offering no specific timeline. I am persuaded that focus and coordination are today's leading challenges, rather than questions of technology per se, yet informed judgments regarding the potential pace of progress will require an understanding of the technological terrain to be crossed.

As we've seen, advances in AP fabrication technologies will directly enable further advances as new tools are applied to develop new tools. These advances will enable faster cycles of design, fabrication, and testing that lead in turn to further advances and yet faster cycles. Dates cannot be predicted, but the pattern seems clear. We can expect to see an accelerating upward spiral of capabilities, with APM-level technologies as a natural destination.

The Potential for Surprisingly Fast Deployment

On the path from systems-level design to actual use, a new product faces delays in detailed design, prototyping, testing, and redesign, then production engineering, testing, and scale-up, and finally, distribution and market adoption. None of these steps is immune to change.

With APM-level capabilities in play, design times for better-than-conventional products can be short and yet yield remarkable results. For example, even partial upgrades of existing products that involve replacing structural components with materials that are lighter, stronger, and lower in cost can offer striking advantages. If a business today could deliver replacements for products already in use, but at lower cost and with superior performance by a few key metrics (vehicles with half the mass, electronic systems with ten thousand times greater capacity), one would expect to see rapid replacement of competing products along with the collapse of the supply chains behind them.

Looking beyond limited upgrades, it's important to keep in mind that cycles of product improvement (and replacement) can be swift with an APM production infrastructure; the delays of prototyping, production engineering, and plant construction largely disappear, and production itself can be both fast and scalable. Further, for products adapted to decentralized APM-based production, distribution need not involve shipping and can more nearly resemble an Internet download.

Deep cuts in costs, advances in performance, swift satisfaction of demand, and displacement of existing supply chains—these characteristics of APM-based production apply to industrial products of every kind. In typical markets, the spontaneous dynamics of supply and demand would drive change at unprecedented speed. Potential successes by a host of metrics—radical improvements in labor and capital productivity, production cost, product performance, energy conservation, carbon emissions, and overall environmental impact—all bring in their wake the potential for positive yet disruptive change.

DISRUPTIVE SOLUTIONS TO GLOBAL PROBLEMS

Physics and engineering point to new capabilities; history and common sense suggest at least some of the unintended consequences to be enjoyed, endured, managed, or forestalled. The prospects include solutions to profound global problems that are closely linked to a range of unexpected challenges.

The Slow-Motion Crisis of Industrial Civilization

Without sustainability, greater abundance would mean a shorter path to failure in an increasingly divided and fragile world. With sustainability, the twenty-first century offers more hope. The previous chapter explored applications of APM-level technologies to a wide range of human needs, and their implications for resource requirements and economic sustainability. How does this potential match up with emerging global problems?

We face a great, slow-motion crisis, the collision between industrial civilization and the limits of Earth. Recent trends and current conditions have injected a sense of urgency. In what may be the clearest marker of emerging resource problems, price trends that had been favorable through the end of the twentieth century have worsened in recent years. After decades of decline, a wide range of industrial raw material prices are increasing. Some regions have always faced scarce water supplies, but those regions are growing. The energy crisis of the 1970s gave way to freer supplies, but petroleum output has recently flattened while prices have spiked to unprecedented heights. Food prices fell during much of the twentieth century, but have risen sharply since 2000, and in recent years high global grain prices have sparked riots. The great slow-motion crisis seems real enough.

These concrete trends give credence to earlier projections. The book that had alarmed me in the 1970s, *Limits to Growth*, had projected smooth growth through the end of the twentieth century, with many possible paths to disaster in the decades beyond. The turning of trends lends support to those century-long projections made forty years ago and the consequences of overloading the Earth with greenhouse gases fits one of the templates.

Regarding current economic development and human well-being, the greatest single positive trend has been the rapid economic rise of China and the slower yet great rise of India. This enormous human progress, however, is widely thought to be unsustainable within the bounds of the Earth's limits and projected industrial technologies. Tight resource supplies today and the CO_2 problem point to a collision, not only between global civilization and the limits of the Earth, but also between rising powers and the powers of the status quo. In round numbers, if China and India were to consume resources at the US per capita level, world resource consumption would more than double and accelerate the rise of CO_2 levels. As one might expect, US military planning documents express concern that resource scarcity could bring conflict in the decades ahead.

Solutions for Material Economic Development

As we've seen, APM-based production technologies can draw on common resources of raw materials and energy to enable material abundance in ways that are radical on multiple dimensions. One radical aspect is the potential decoupling of material abundance from its traditional societal preconditions.

Economic development and human development are intertwined, each affecting the other. Economic development lifts populations out of hunger and rural poverty toward lives free of urgent want, enabling children to devote time to education and enabling society to support educators. Advances in human development, in turn, enable societies to join today's global industrial society and rise further.

APM-based production, however, can enable the material plenty needed for human development while breaking the link between human development and material economic progress. Cell phones aid development, yet to use these devices requires neither advanced training nor a strong societal infrastructure of law, savings, investment, or education, and APM-based production has similar characteristics. Image files today will be joined by product files tomorrow. Today one can produce an image of the Mona Lisa without being able to draw a good circle; tomorrow one will be able to produce a display screen without knowing how to manufacture a wire.

Today, lack of material development and human development can reinforce one another in a circle that perpetuates poverty. A new way to make things can break that circle, with unforeseeable outcomes.

The same pattern holds in the developed world, representing another break from past patterns. Advances in material production have required complex new technological infrastructures—the production of automobiles, for example, requires an intricate supply chain both for the products themselves and for building the factories that produce them. Advances have increased complexity and spun a web of interdependent economies that grows denser year by year.

APM-based production will encapsulate complexity in a form that users have no need to examine. The complexity of a cell phone's computer chip is enormous, yet self-contained; its complexity is beyond the full comprehension of any person alive, yet to use it requires little skill or investment in learning. One need not even know that chips exist. Here once again, parallels between APM and digital systems provide a good model where parallels with industrial systems fail.

The potential material standard of living enabled by APM-level technologies is perhaps best left to imagination for now. The base level stands somewhere above a world-wide abundant supply of the best of every kind of product manufactured today.

Solutions for Sustainability and Resilience

Culling reports from a range of international and non-governmental organizations yields a set of requirements for sustainable global development. One group of these requirements centers on resources and production:

- Conserving material resources
- Providing ample energy, water, and food
- Reducing poverty and elevating the standard of living

As we've seen, advanced atomically precise fabrication leads naturally to a shift in resource demand, both in quantity and composition. Earth-abundant raw materials suffice for making products superior to those based on scarcer materials, like copper and zinc. A fall in demand will make most scarce raw materials no longer an economic concern, while prospective improvements in extraction and recycling push in the same direction.

Likewise, low-cost production of photovoltaics can enable production of abundant solar electric power; new AP technologies for interconverting electrical and chemical energy (stored as hydrogen or hydrocarbons) can enable efficient energy storage and fuel production, overcoming problems with solar energy posed by night, weather, and geography.

With abundant energy and a radical reduction in the cost of capital goods, desalination and long-range water transportation become practical means of providing fresh water on a scale that can sustain large-scale agriculture. Similarly, low-cost energy, equipment, and structures can enable low-cost enclosed agriculture, multiplying yields while conserving water and retaining nutrients (nitrogen, phosphorus) that now flow from open fields, polluting streams, lakes, and rivers—even creating dead zones in the sea itself.

Perhaps even more important is a natural, qualitative result of expanding capabilities: unprecedented resilience. Resilience in a material sense amounts to tolerance for external shocks from the natural and human worlds. With enclosed agriculture, drought is no longer a concern, and with abundant energy and cooling machinery heat waves need not result in deaths. Even the literal shocks of earthquakes become lethal largely though building collapse, where the relative fragility of construction is a consequence of the cost of more robust structures, which in turn is a consequence of today's costs of production. The same can be said of vulnerability to floods, storms, and fires.

Many other shocks (including some caused by human action) also take a straightforward material form. Villages, regions, and nations can suffer from failures of food, water, electric power, or fuel supplies or from a fall in demand for local products, cutting off access to funds for buying necessities. Physical and economic decoupling can mitigate these problems by enabling local production. Short supply chains need not cross oceans or continents, while preferences for fresh, locally produced food, enabled by enclosed agriculture, would tend to decouple communities from remote sources of food. In energy, decoupling can alleviate dependence on overseas oil production and freedom of passage for oil tankers traveling through the Strait of Hormuz. Because production of physical goods can be as widely disseminated as information technologies are today, local supplies of these goods can be freed from dependence on exports and imports.

Of course, the mere existence of APM-level technologies guarantees neither resilience nor widespread abundance. New capabilities can, however, transform these objectives from costly and perhaps impossible dreams into practical, low-cost, nearly spontaneous options—though outcomes will, as always, depend on choices and the course of events.

Solutions for the Environment

Beyond resource requirements for sustainability—raw materials, energy, water, food—are requirements for protecting the natural environment. The reports mentioned above commonly highlight three:

- Reducing pressure on ecosystems
- Reducing emissions of toxic chemicals
- Reducing emissions of greenhouse gases

The spread of agriculture has placed the ultimate pressure on ecosystems across much of the Earth, the prospect of complete obliteration. Fertile land is biologically productive land and biologically productive land—usually the former home of dense, rich life—is precisely the land where ecosystems have most quickly disappeared under the plow; over the past several generations more of the fabric of life has vanished than anyone living remembers. More is destroyed every day.

Opportunities to multiply yield through enclosed agriculture can slow or even reverse this trend. The incentive to farm new land will drop and in the natural course of events much farmland would be freed. Modest incentives could be effective in freeing yet more land, not just in marginal grasslands, but in lands fertile enough to have once sustained forests. Decoupling agriculture from traditional constraints can radically reduce pressures on nature. There seems to be no other way of accomplishing this goal.

As for toxic emissions, APM enables precise control of material flows and will reduce demand for scarce, toxic elements. APM processes need not produce or emit toxic chemical wastes, and a changed pattern of demand will tend to leave heavy metals in the mine.

Does this mean that APM-based production will, by itself, solve any of these environmental problems? The answer, of course, is No. What it can do, however, is enable more effective solutions, with both greater results and lower cost. Because pressure yields greater results when there is less cause for resistance, this shift in the balance of costs and benefits can amplify the effects of environmental action.

Solving the CO_2 Problem

There's a widespread misunderstanding of the CO_2 greenhouse-gas problem even among those who see the problem as urgent. Although the consequences of rising CO_2 levels are widely recognized, even deep reductions in CO_2 emissions would fail to avert increasing effects because CO_2 doesn't behave like familiar pollutants—it persists.

In light of persistent controversies, a few simple facts bear repeating. CO_2 increases the amount of heat retained by the Earth's climate machine, with effects amplified by increases in water vapor. Large effects are directly observable by the general public—disappearing Arctic ice, earlier springs, and the poleward shift of temperate-zone species. Regarding expert projections, computational climate simulations have defects, but because their errors can go in either direction, uncertainties aren't reassuring (the pace of sea ice melting, for example, far exceeds projections). Likewise, although past climate fluctuations aren't well understood, this, too, is far from reassuring: Signs of spontaneous, unexplained instability in the climate system give reason to worry about the consequences of giving it a kick.

What makes climate change an unusually difficult problem (aside from the scale of its effects and the momentum of fossil fuel consumption) is the long-term persistence of CO_2 in the atmosphere. Levels of atmospheric pollutants like soot or SO_2 closely track the rate of emissions. Turn off the source and the pollutant soon washes out of the atmosphere. Even chlorofluorocarbons and methane have lifetimes measured in a small span of years.

CO_2 is different. It exits the atmosphere on a timescale best measured in decades or centuries. Rain doesn't scrub CO_2 from the atmosphere; instead, CO_2 is slowly absorbed, primarily by the seas.* At any moment, CO_2 levels are only slightly elevated by last year's emissions and are instead the cumulative result of emissions that span the industrial era. Deep cuts in emissions wouldn't lower CO_2 levels in proportion, but would instead merely slow down the rate of their rise. The system behaves more like a bathtub with a drain that's almost closed: If the tap is left open, the tub will eventually overflow.

---

* While it is true that plants capture CO_2, this has little net effect, since the return of CO_2 to the atmosphere by animals, bacteria, and fungi roughly balances photosynthetic uptake, making the biosphere approximately carbon neutral.

Studies of Earth's chemical and geophysical cycles indicate that temperatures and CO_2 levels would remain high for centuries even if emissions were cut to zero today; thus, it seems that only atmospheric carbon capture technologies can provide a large enough drain to lower CO_2 levels quickly and deeply. As I discussed in the previous chapter, to accomplish this task—to collect and compress three trillion tons of CO_2 in the span of a decade or so—would require energy resources and equipment that would strain or exceed the capacity of our industrial civilization, yet are well within reach of a world with low-cost APM-based production.

Needless to say, to rely on this prospect today seems unwise.

Dimensions of Disruption

Deep, rapid change in human affairs can mobilize forces that are more predictable than their effects, responses, or eventual outcomes. Here I'd like to outline a few general prospects, not as predictions, but as potential effects of disruptive forces—including effects that could perhaps be moderated by well-chosen responses.

There's great value in asking not only appropriate questions, but also an appropriate range. It seems that multiple radical capabilities will arrive bundled together. Each will change the context of the rest, often in radical ways, and the questions they raise must therefore be considered together. The discussion that follows attempts to draw some of the connections, but I leave it to others to consider a yet broader range of questions and to delve more deeply into potential answers and how they fit together.

Two kinds of questions arise in each area of concern, first *What consequences can we reasonably anticipate from new capabilities?* which is to say, consequences of newly accessible capabilities, whether they are actualized, left fallow, or forestalled; and then the closely related question, *What are the likely consequences of anticipation itself?* which is to say, consequences of human expectations regarding technological prospects, whether or not these expectations are well-grounded and whether or not the responses make practical sense.

Expectations shape perceptions and actions, sometimes long in advance of events. As progress toward APM continues, we can expect related expectations to grow and to gain increasing influence. History suggests that realistic expectations—and realistic assessments of uncertainty—are far from assured, and yet expectations inevitably shape outcomes. In considering the impacts of prospective products, production technologies, changes in asset values, and environmental applications, expectations are central and can take form at a much earlier date.

The Scope of Potential Products

The potential for a vastly expanded scope of production raises concerns regarding what might be produced. Potential products of advanced fabrication include kinds that today are subject to strict regulation, such as weapon systems, drugs, explosives, toxic chemicals, and so on. Even without including anything exotic, these examples provide sufficient reasons to want—and expect—restrictions on the range of APM products.

From an engineering perspective, it is straightforward to design APM systems that can produce only a restricted range of products. Implementing restrictions on core technologies—the machines used to make different and better machines—is a very different challenge, however, one that calls for substantial transparency and oversight during late-stage development. This, in turn, calls for implementing programs, policies, and institutions in a process made more complicated by competitive forces, both commercial and international.

Thus, APM-based production technologies will present policy challenges well before they emerge. Although different in kind, these challenges follow from the physical character of AP fabrication as directly as the challenges of nuclear weapons technology followed from the physical character of fission.

The Scale and Organization of Production

Together with scope, the sheer scale of production capacity ensures disruptive potential. We've seen how the characteristics of APM-based production can shorten the path from design to initial products and from there to rapid scale-up and market dominance driven by unprecedented advantages in cost and performance. "Rapid" here is a relative measure, faster for individual products than for whole industries, but the natural dynamics of competition extend to that scale. Economic change that renders entire industries and their supply chains obsolete—and not one, but many industries—would quickly reach macroeconomic proportions, requiring an economic analysis that takes account of job losses in multiple sectors, widespread declines in the real prices of goods, and knock-on effects on purchasing power and demand.

Perversely, the supply-side potential of radical abundance could lead to widespread and radical want. Whether from the perspective of corporations or people, creative economic destruction would be radical in both its creative and destructive dimensions.

History suggests the likelihood that large-scale interventions will be undertaken to moderate both the pace of economic change and the human costs of economic disruption. To be effective, however, interventions of this kind will require coordination on the scale of nations and the global economy.

I leave it to readers to consider potential winner-take-all outcomes and whether they would want to live in the aftermath, almost certainly not sitting on top. As for what disproportionate winners might claim is their right, by the rules of one game or another, every such claim must be placed in perspective. Each advance in technology adds only an increment to a tower of achievement built over millennia by millions of human hands and minds, and that tower is, in both practice and age-old principle, the common heritage of humankind. Each generation contributes, and no living person can rightly claim ownership of more than a tiny and transient part.

Economic Revaluation

An economy transformed by APM-based production would undergo radical revaluation—which is to say, devaluation—of many physical assets, including ore deposits, oil fields, and durable capital goods (power plants, for example). These assets differ from common products because their net present value depends on developments well beyond the usual short-term horizons, and assessments of value are therefore sensitive to long-term expectations. Regarding the examples above, an anticipated fall in demand for metals and petroleum reduces the expected value of ore deposits and oil fields, tending to reduce resource prices and increase consumption. Likewise, an expectation that coal-fired power plants will become obsolete reduces their expected net present values, discouraging expansions in power plant capacity.

Thus, changing expectations regarding the pace and implications of APM-level technology development can change assessments of the risks and net present values of a range of assets. In addition, however, expectations regarding near-term asset prices are subject to a second-order consideration, the expected changes in others' expectations and their consequent asset valuations.

Environmental Prospects

Prospects for a radical increase in the quantity of production raise concerns because even inherently clean, low-footprint technologies can enable greater encroachment on nature. Favorable outcomes will depend on cultural values and on cooperative measures based on incentives and regulation.

As with revaluation of resources and capital, long-term expectations can have more immediate effects—not just on market valuations, but also on the balance of concerns that sway political action. For example, the prospect of reversing the rise in CO_2 levels at some indefinite time in the future may affect opinions regarding measures to curb emissions today. In light of uncertain technology development timelines—and the persistence of irreversible environmental effects—the risks and costs of tampering with the Earth's climate system remain all too real. Neither fast nor slow development can be assured, and neither should be assumed.

Regarding the effects of anticipation on environmental decisions, expectations of fast timelines cut in both directions. The scenarios that deliver early opportunities for remediation are also scenarios that undercut the value of investments in large, durable, destructive projects. The case for eventual harm reduction sometimes tilts the balance toward avoiding the harm from the start.

COHERENT EXPECTATIONS, INTERESTS, AND RESPONSES

Again, sound thinking about the prospects ahead requires an effort to consider the full range of interlocking concerns. Because the technology of APM-level production has so many ramifications, a coherent view of potential futures must not assume a few and ignore the rest—not imagine bits and pieces of change against the background of conventional expectations.

For example, radical advances in medicine promise to extend healthy lives, and thereby invalidate current demographic projections. What does this imply for future retirement costs? Changing one number in a spreadsheet built on standard projections would suggest outcomes that make little sense in a future in which lower costs of production lighten the burden of supporting a non-working population, while the prospect of extending *healthy* lives invalidates conventional expectations of chronic illness and high health care costs. Thus, the implications of APM-level technologies for medicine, lifespan, health, and economics must be considered together to form a coherent picture.

Within the sphere of economics, coherence is also essential. If considered against the background of a conventional future, one might try to project impacts on employment based on experience in a particular industrial sector (steel, for example). To be realistic, however, an economic analysis must be framed more broadly as part of a systemic transition. In a world with APM-level technologies, the range of challenges and options will be unlike past experience. Relying too much on analogies to past events will suggest familiar but unrealistic consequences while omitting a new, unfamiliar range of potential responses.

Once again, the ongoing revolution in information technologies offers analogies. Both the Information Revolution and the APM Revolution bring in their wake unprecedented ranges of capabilities based on a general-purpose technology, and each in its sphere brings a kind of radical abundance. We've seen the emergence of a gift economy in digital products such as software, text, images, and video; the natural course of events would see this pattern extend to APM product-design files, leading (aside from the cost of input materials) to a gift economy in physical objects (but within what mandated constraints?). Considering both similarities and contrasts between the two revolutions can help to build a more robust conceptual framework.

Regarding specific problems (a plunge in demand for steel, for example), it's natural to worry that they might be neglected, simply out of distraction and inertia. In a world in which coherent scenarios have traction, however, specific problems will often be framed as instances of a broader, generic, high-profile problem—steel, after all, won't be a special case of falling demand.

Likewise, concerns about particular hazards of unconstrained applications of APM—this material, that device—won't arise in isolation. In a world in which coherent scenarios have traction, specific concerns about the abuse of APM-level technologies will likewise be framed as aspects of a more general problem of constraining APM applications. Provided that attention centers on coherent scenarios and coherent responses, specific, important concerns of this kind will best be addressed as part of a broader agenda; if incoherent scenarios dominate the conversation, however, particular concerns are more likely to be lost in the confusion and would be less likely to find good solutions. There's little to recommend fragmentary responses to a civilization-scale transformation.

Incoherent scenarios foster incoherent responses that create or perpetuate needless risks. In particular, the prospect of radical abundance, if well understood, can mitigate some causes of conflict. In competitions for markets, resources, and the power to secure them, and in struggles that pit industrial production against environmental protection, participants find themselves in a near-zero-sum situation, where gains by one side entail loss by the other, and where each side might expect to gain from the other's errors. In each of these areas, however, the prospect of radical abundance offers the potential for gains on all sides, while habits of struggle and opposition, if pursued with unchecked momentum, would bring needless conflict and opportunity costs. In the military sphere, in particular, seemingly rational decisions, if informed by incoherent scenarios, could risk disastrous outcomes in pursuit of illusory gains.

A REALISTIC, shared understanding must take account of both knowledge and uncertainty. Physics provides a starting point, a basis for well-founded knowledge of some aspects of technological potential, and this potential, in turn, presents a set of partially predictable opportunities and challenges.

Beyond this threshold (and beyond engineering calculation), prospects can be considered on a rational basis, yet confidence in particular outcomes is apt to be misplaced. Questions of future technology timelines, winners and losers, perceptions, opinions, politics, events, responses, policies, economic impacts and societal outcomes—in all these areas, predictions are dubious, yet better understanding can lead to better decisions.

One form of future-oriented knowledge, however, can be robust: knowledge of the scope of uncertainty itself, of the extent of questions that have no reliable answers. Uncertainty can be important to recognize and examine because it has consequences for making rational, risk-averse choices. There is no paradox in seeking better knowledge of uncertainty itself.

Making sound, risk-averse decisions commonly requires conservative assumptions, but uncertainty about APM development timelines gives "conservative" two contrasting meanings. When contemplating potential solutions to problems (economic, medical, environmental, and so on) it's conservative to assume that long delays will arise on every path taken toward radical abundance; when contemplating potential new problems, however, it's conservative to assume that at least one path will be surprisingly short.

CHAPTER 17

Security for an Unconventional Future

FROM HAMMERED IRON to digital chips, disruptive foundational technologies have always found military use, and APM-level technologies will surely be no exception. Unlike previous disruptive technologies, however, APM-level technologies promise to change basic rules and the context of the game itself. Because these technologies will enable a radical abundance of both conventional goods and security systems, any strategy for managing APM-level capabilities must take account of both sides of the potential conjunction of material wealth and the technologies of power in an unconventional future.

The advent of disruptive technologies of power need not destabilize military and societal order. These technologies, if applied with an adequate measure of wisdom, can provide new means of moderating conflict as well as a means of protecting benign, accountable forms of social order that increase both liberty and security.

Nonetheless, these same technologies could easily spur an arms race that precipitates war, or could empower secretive surveillance regimes that threaten even their would-be masters. Courses of action that increase these risks would pose needless and incalculable threats to both states and individuals, because grabs for unaccountable power are a simpler, unilateral, age-old response, while developing stable and accountable governance requires greater thought, consensus, coordination, and institution building.

To enable a well-grounded discussion of risks and opportunities, we must explore the potential capabilities offered by APM-level security technologies, and in the context of a broader APM transition. Despite the limits of what can be seen today, enough is in view to show some of the forces that must be considered in order to frame coherent questions about a future outside the bounds of conventional expectations.

DISRUPTIVE POTENTIAL: MILITARY ASYMMETRIES

Competitive technology development efforts rarely proceed in synchrony and the history of technological competition, whether in nuclear weapons, satellites, computer chips, or stealth aircraft, shows that competitors typically reach technology thresholds at times that differ by years or more. What is different in the realm of APM-based technology is the potential speed of advanced-stage development and scale up, which could dramatically heighten the resulting disparities in capabilities. It is therefore easy to envision scenarios in which a modest degree of asynchrony, measured in months or less, could swiftly lead to radical military asymmetries even with relatively shallow exploitation of the overall potential of APM-level technologies.

Consider more-or-less conventional aerospace systems, where APM-based production can enable a one thousand–fold reduction in cost per unit along with improvements in performance (payload and range, sensors and computation, etc.). With APM-based production, $10 billion can supply enough resources to build ten million tons of high-performance aerospace vehicles and manufacturing facilities with enough capacity to accomplish this in a matter of days. For comparison, note that ten million tons would suffice to build ten million conventional cruise missiles, a thousand times as many as in the current US arsenal.

Potential quantitative and qualitative asymmetries on this scale could enable overwhelming military dominance. If development were to proceed in an unconstrained, competitive framework, it seems that an arms race would be inevitable. This would shape the entire context of APM technology development—perceptions of the technologies and their significance, the balance of applications to arms vs. economic development, climate change, medicine, and so on. Moreover, secrecy would increase uncertainty on all sides, because, in this sphere of technology, relative positions in a secretive arms race would be uncommonly difficult to monitor and breakout applications would be hard to anticipate.

These circumstances call for a careful reassessment of national interests and policy options. Attention to both technological potential and patterns of history can help to inform judgment.

Disruptive Potential: Non-lethal Force

APM-level production capacity could be applied to build larger arsenals of lethal weapons like missiles and bombs, yet refined technologies and radically lower costs will enable options of a different kind. In particular, a shift in the relative advantages of non-lethal weapons will enable a profound change in how force is applied.

In the 1950s, thermonuclear weapons marked a triumph of lethality over precision. Thermonuclear weapons enabled nations to wage war at the lowest cost per kill ever seen, but without the ability to hit small, specific targets without slaughtering civilian populations within range of blast or lethal fallout. It is telling that such weapons have been ready to use at a moment's notice for more than half a century, yet have served only as threats, while new weapons systems have trended toward greater precision, less overall lethality, and more facile use.

The weapons that make news today—miniature aircraft, missiles, and guided bullets; applications of surveillance systems and drones—all point to an interest in applying lethal force from a distance, yet with the precision of sniper fire.

It is important to remember, however, that coercion—not killing—is the usual purpose of war. As Sun Tzu said, "Supreme excellence consists in breaking the enemy's resistance without fighting," and in his best-known quotation Clausewitz described war as "merely the continuation of politics by other means." Accordingly, the primary use of weapons is coercion—and although killing has to date been more economical, the economics of advanced technologies and low-cost production will make non-lethal weapons more competitive.

Imagine a swarm of unmanned drones—built at low cost and in enormous numbers—pitted against conventional air power. Air defenses could be saturated if there were sufficient drones—they could simply absorb hits until a defender's munitions ran out. A sufficient number of drones kept on station with suitable sensors and precision-guided munitions could punch holes in the wings of aircraft before they take flight, destroy ballistic missiles in early boost phase, detect and destroy cruise missiles moments after launch, and so on (all with little lethality). The same strategies could be applied to suppress shipborne weapons, and with a dense, low-cost sensor network the seas would give no refuge to submarines. In effect, wars would be won or lost before they had even begun. Precise destruction of non-military targets could likewise pressure or incapacitate an entire society. War would become less costly in terms of both money and casualties (zero on one side, low on the other), and hence wars would be relatively easy to initiate and win.

Here we see the logical outcome of today's increasing reliance on unmanned aircraft. Weapons that place fewer lives at risk can encourage more warlike policies; widespread US use of drones, though lethal, already illustrates this principle. Abundant, affordable, non-lethal, remotely operated weapons would go further, severing the link between making war and killing people. With reduced moral qualms, the threshold for action would fall. Because an opponent facing this prospect would therefore have less reason to expect restraint, this side of the equation strengthens the logic of preemption.

Disruptive Potential: The Civil Society Dimension

The potential of APM-level production capacity doesn't end with the military. Already, advances in nanoscale electronics have brought low-cost video imaging, storage, and data transmission to pocket devices and the increasing ubiquity of such features reflects the plunging cost of compact sensors, computers, and telecommunications systems. These technologies have been used to good effect by citizens, for example, in applying pressure to central authorities by publicizing state responses to protests in Tunisia and Egypt, which raised the political cost of violent suppression. Advances in non-AP semiconductor nanotechnologies have driven these trends. Ongoing progress will carry them further.

Ordinary citizens are not the only beneficiaries, of course. The cheaper and more compact these technologies become, the lower the cost for governments to implement dense surveillance networks. In parallel, as surveillance hardware improves, so does software for image recognition—pinpointing objects, people, and faces with startling accuracy even today. And indeed, here in Great Britain I live in a nation that has the highest density of surveillance cameras in the world, both private and installed by the state; meanwhile, deployment of surveillance networks is rolling forward in both the United States and China.

Even current trends in technology raise pressing questions about the spread of surveillance and drone technologies. In terms of capabilities, the door is now open to developing and deploying autonomous systems that can observe, recognize, and report persons and events, and can incapacitate, mark, kill, or harass people, whether by program or human intervention, all the while collecting evidence for courts—a crucial consideration for a society that demands lawful, accountable governance.

Coherent expectations of an APM-driven future must take account of these trends, both before and during the APM technology transition. These trends raise concerns and create possibilities that deserve deeper consideration. I am persuaded that more can be gained by advocating and crafting institutional frameworks for managing new capabilities than can be gained through more simplistic advocacy or opposition to the basic capabilities themselves.

THE PARADOX OF RADICAL DOMESTIC SECURITY

Advances in technology that seem to drive change in a single direction can lead to results that reverse directions in midstream. This seems to be true of security systems and their implications for liberty and governance, and history provides parallel examples.

As noted above, advances in military technologies led first to an increase, then a decrease, in the lethality of war. Retail killing with axes and muskets gave way to wholesale killing with machine guns and nuclear explosives, yet wholesale killing is now giving way to retail-scale killing by covert non-state actors and state-employed drones.

Looking further back in history, advances in the social technologies of governance similarly led to a decrease, then an increase, in personal liberty. Early state institutions added a layer of control on top of agrarian societies, with arbitrary power over life and death for their subjects. State institutions nonetheless evolved over time, increasing in power and potential lethality, yet (in some places and at some times) this power has supported societies governed by law and enabling concrete liberty—personal safety, personal choice, and constraints on coercion—to a degree beyond prior human experience.

Security technologies could follow the same paradoxical pattern. Although advances in surveillance capabilities have demonstrated their potential to erode liberty, they could also be applied with an opposite effect, increasing safety and liberty alike. In the United States, discussions of domestic security now center on potential terrorist attacks, with disputes about acceptable risks and the effectiveness and acceptability of security measures implemented in secrecy and with limited oversight. Debates hinge on whether the infringements on personal liberties are adequately offset by improvements in security, and whether the watchers themselves must be allowed to work in secrecy in order to shore up weaknesses in security systems.

How might this calculation change with more effective, intensive surveillance? Looking toward the radical end of potential capabilities, imagine a world in which substantial terrorist plots can readily be detected and thwarted. An attack would be almost impossible to plan in secrecy, and then (if planned) would be almost impossible to organize, and then (if organized) almost impossible to execute. The resulting world could be safer, yet it would seem to be the kind of world that civil liberties advocates fear most.

But need this be true?

If the problem of defense were effectively solved, other concerns would then weigh more heavily. With thoroughly effective means of collecting evidence, for example, open judicial proceedings should reliably lead to appropriate convictions. In moving toward such a world, the value of keeping evidence and its sources secret would dwindle, while the traditional, powerful value of open judicial proceedings—as a check on abuse of power, as a shield to protect liberty under law—would gain greater weight.

Can the "long war" against terrorism be won? Technological trends suggest that the eventual answer will be Yes. If so, then the longer-term challenge won't be to protect populations against attack by outsiders, but to protect the fabric of civil societies against destruction from within, to avoid the entrenchment of an opaque and unaccountable security apparatus of the sort that history suggests threatens everyone, not only society as a whole, but also those who imagine that they can control it.

If liberty, accountability, and effective security aren't necessarily at odds, then it seems important to explore how they might be reconciled. I see an urgent need for discussions that reflect widespread agreement about what would be good, in principle, if the costs were low enough. For example, personal safety, national security, and traditional freedoms are widely valued despite heated arguments regarding how much of one must be sacrificed in order to gain more of the other. Typically, neither side of a dispute rejects what the other side values more highly, yet because they weigh shared values differently, policies favored by one side may be opposed by the other. With the prospect of radical abundance, however, trade-offs can be radically improved, offering options that can improve outcomes by several criteria simultaneously. If recognized, the prospect of broadly superior options like these could (and might) prompt all parties to both expect and get more of what they value, and to seek greater consensus.*

---

* Likewise, as noted before, industrial technologies have forced trade-offs that pit environmentalists against development advocates, even though neither camp favors poverty or environmental destruction. Here again, looking forward, new options can enable across-the-board gains.

Options for managing information-intensive security in a civil society are surprisingly broad, yet the range of potential patterns of information collection, access, response, and accountability has scarcely been explored in public discussion, even in outline. When exploring potential solutions to problems, institutions, law, digital technologies, and procedural rules must be considered together—and in creative, culturally appropriate combinations—or much will be missed. I am persuaded that these means, in combination, could be applied to structure transparency and force in surprisingly effective and acceptable ways that could foster personal and national security together with traditional freedoms in open, dynamic civil societies.

MORE CLOSELY ALIGNED NATIONAL INTERESTS

We've explored some of the ways in which APM-level capabilities can change security technologies and have seen the need to explore new trade-offs and choices in the domestic sphere. In the international sphere, APM-level technologies will likewise bring new trade-offs and choices, changing vital national interests with profound security implications. In particular, ensuring access to markets and resources loses importance, while restricting access to APM-level technologies becomes critical.

Less Need to Compete for Markets and Natural Resources

First and foremost, broad international access to appropriate APM-level production capabilities would decrease pressures to compete for access to markets and natural resources simply because there can be no vital interest in resources that are no longer scarce or important, nor a vital interest in export markets once imports and trade balances are no longer essential to material well-being.

Today, a nation would be crippled without access to oil supplies. Access to suitable APM-level production capabilities would enable rapid, economical energy independence. Oil imports would no longer be critical.

The same can be said of industrial raw materials, components, and products. Widespread access to key APM-level technologies would make autarky practical. Where food is concerned, the productivity of enclosed agriculture is high enough that even a nation with the population density of Singapore could feed itself from domestic production. Indeed, it is hard to find an external resource that would continue to be a vital national interest, a circumstance that would blunt an age-old cause of international conflict.

Strong, Shared Interests in Constraining Non-state Actors

As we've seen, APM-based production systems can be designed to manufacture only particular ranges of products; these can include just a few consumer goods or a full spectrum of goods, energy systems, medical products, and the like as well as the components needed to expand production capacity, all without providing access to products that would present a security risk. Machines built with specialized capabilities have inherent constraints, and causes for concern regarding access to non-military capabilities are essentially economic, environmental, and (broadly speaking) regulatory rather than strategic in nature.

By contrast, unconstrained access to an unconstrained range of APM-level technologies would place unpredictable capabilities in the hands of hostile non-state actors, leading to unacceptable and unpredictable risks. Preventing unconstrained applications of APM-level production capabilities is thus a vital national interest of the highest order, one shared by all nations, and a potential driver for increased worldwide cooperation in addressing APM-linked security concerns.

This concern gives nations a vital interest in the domestic security of other nations, a shared interest in establishing effective global restrictions that entails an interest in the capability and will of other nations to restrict the dissemination of APM-level technologies within their borders.

Fewer Fundamental Reasons for International Conflict

History offers no close parallels to the present situation. Throughout historical times (and before), competition for natural resources has shaped societies, forcing confrontations with rivals and fostering a propensity for war. From hunting grounds to oilfields, control of territories and access to scarce resources have been critical to strength and survival. Today, however, we face the prospect of an era in which resource competition becomes comparatively unimportant.

Today, resource competition continues to be a leading cause of international tension; conflict over control of the East and South China Seas, for example, is premised on the assumption that mid-twenty-first-century nations will still have an interest in undersea oilfields. Because the emergence of APM-level technologies will make most resource-centered concerns obsolete, plans for an unconventional future cannot count access to natural resources as vital or even substantial national interests. To expect both APM-level technologies and a continuing struggle for resources would be incoherent, fostering needlessly risky plans for an illusory future.

Aside from material concerns, the twentieth century was marked by struggles that were at least nominally about ideology, yet today, at the level of major powers, ideological struggles have faded into the background and are no longer discussed as plausible causes for war.

The remaining external threats to vital national interests have a circular character, in which responses to threats to security give rise to reciprocal threats in a cycle that can become autonomous when past events, not fundamental interests, shape the dynamics. During the United States/Soviet arms race, for example, military threats were disproportionate to conflicts of material interests—the United States and Soviet Union scarcely competed for markets or natural resources, yet wrestled for power while counting warheads and contemplating mutual assured destruction. By contrast, the United States and Europe compete in every dimension of economic affairs, yet they show no perceptible concerns regarding imbalances in military power: the United States, Britain, and France have stockpiles of thermonuclear weapons of greatly different sizes (thousands vs. hundreds), yet without a recent history of reciprocal threats, these disparities are not regarded as important.

Meanwhile, between the United States and China, a history of conflict and past military clashes has fostered ongoing military confrontation despite close economic ties. It is hard to imagine an arms race between the

United States and Europe, yet easy to imagine one between the United States and China; indeed, in a highly asymmetrical way, a race is already in progress.

Despite the enduring importance of military strength and arms development, an unconstrained arms race centered on APM-level technologies would pose grave and unnecessary risks. The characteristics both of the technologies and of potential development paths present unfamiliar challenges in a strategic context. To avoid blundering into an untenable situation, both risks and alternative strategies must be reconsidered, starting with the most fundamental concerns.

UNCERTAINTIES, RISKS, AND WAYS FORWARD

Profound changes in national interests will call for a ground-up review of grand strategy. Means and ends, risks and opportunities, the future self-perceived interests of today's strategic competitors—none of these can be taken for granted. A radically unconventional future cannot be accommodated within the framework of plans made for a different world.

Looking forward, some kinds of knowledge can be established with confidence; others cannot. The physical potential for APM-level technologies can be explored by means of engineering methodologies, and lower bounds on potential technological capabilities can be established with confidence, yet upper bounds often remain speculative. In estimating development timelines, physical calculations offer little help, yet the status and prospects for accelerating development of AP technologies argue strongly for the emergence of APM-level technologies in a time frame that calls for contingency planning. Both knowledge and uncertainty call for a response.

Embracing Irreducible Uncertainties

In financial affairs, markets are expected to behave unexpectedly, and market risks are central to judgment. Thus, although investors inform their decisions with knowledge in the conventional sense, they also embrace irreducible uncertainty, and explicitly so.

In the years ahead we can expect to see APM-level technologies emerge, followed swiftly by large-scale applications, but because the pace of progress cannot be predicted, neither fast nor slow progress can be safely assumed. Indeed, the specific path of progress cannot be predicted simply because many paths stand open and shortcuts are apt to be found. Looking forward, as concrete technologies emerge, unexpected applications will inevitably appear. And beyond even these technology-centered uncertainties, outcomes will be contingent on expectations, choices, and cascading events that are beyond prediction.

Thus, the coming revolution of radical abundance confronts us with irreducible uncertainties that, as in financial affairs, must stand alongside positive knowledge at the center of decision making. Knowledge and uncertainty interact. Unpredictable prospects often call for hedging and contingency plans; unknowns can give rise to predictable results when *known* unknowns affect others' choices.

Cooperative Strategies for Avoiding Needless Risks

Cooperative development can reduce unknowns, while secretive competition intentionally increases them. With fewer clashing vital interests and shared challenges of managing a turbulent economic and security transition, allowing an arms race to emerge would create a needless and incalculable risk.

For a range of reasons, a course of secretive development would ensure that no party today could have confidence in competitive outcomes and that none could confidently expect to gain confidence even late in the game. The nature of pathway technologies appears to ensure swift progress after advances cross a technology threshold, and reaching that threshold requires neither exotic materials nor a specialized, visible infrastructure. Knowledge is the key to implementation, and knowledge is fluid and can be quickly and invisibly transferred. In the context of a secretive arms race, advances would be too unpredictable, too easy to hide, too difficult to track, and too easy to leak. In that secretive context, for the nations involved—indeed, for the world as a whole—irreducible uncertainties would grow.

As we've seen, at an objective, material level, the APM technology transition can bring better-aligned national interests; it would be ironic if the anticipation of that transition instead increased the risk of war. In light of these prospects, what can be done?

The motivations for secretive military development are clear and real, and although the path they favor is risky, no unilateral policy of restraint can negate the logic that leads there. An alternative path would emphasize openness as a means of reducing uncertainty and risk while focusing on shared problems and the challenges of managing disruptive transitions in domestic and international affairs, with security risks among them. Fortunately, the structure of today's AP research provides a good starting point.

Today, across a wide range of fields, scientific research is strongly international. Multinational scientific collaborations are common and growing—for example, papers in *Science* and *Nature* commonly have multiple authors with affiliations that show crossovers in every direction between the United States, Europe, China, Japan, and Korea. In science, secretive competition is real, but the competitors are most often research groups racing toward achieving results to share with the world through open publication in journals. The international culture of scientific research provides the context for today's most advanced work in AP technologies, a baseline of relative openness and relative transparency. This culture and its institutions could be deepened and extended at levels that range from personal choices to institutional decisions, national policies, and multilateral agreements.

Research objectives shape research cultures. Today, research in relevant AP technologies—a very broad spectrum—contributes to applications in medicine, materials, biotechnology, and process engineering as well as to basic knowledge in fields that include chemistry, materials science, physics, and molecular biology. The most advanced research is typically the furthest from competitive applications and motivated by broad human aspirations. The prospect of solving global problems is among these motivations and naturally fosters cooperative research. And again, these cooperative inclinations among scientists could be deepened and extended as a matter of policy.

If nations—from this beginning—maintain and institutionalize a well-structured transparency around AP technologies, this can provide a favorable environment for further measures to establish trust in one another's development and management of APM-level technologies. This pattern of structured transparency, in turn, could provide a basis for the multilateral, coordinated security efforts that seem critical to international stability.

To follow this path will require a clear vision of APM-centered developments that will give rise to a distinct, unconventional kind of future, a future outside the current range of expectations, a future that isn't at all what we'd planned. Realistic scenarios for an unconventional future must take account of both foreseeable disruptive forces and irreducible uncertainties along with the potential benefits—and perhaps the necessity—of cooperative international responses. No matter how much weight one gives to unconventional scenarios in other policy arenas, nurturing critical AP technologies with these scenarios in mind seems wise as a strategy for reducing risks.

To successfully pursue this path will require more than just habits, inclinations, and informal agreements regarding information sharing. As critical technology thresholds approach, latent military concerns will become increasingly concrete and urgent. This will be a time that tests every aspect of the preparations in place—the degree of shared understanding of uncertainties, challenges, and changed national interests; the degree, extent, and nature of consensus around potential courses of action; the personal and organizational relationships that can turn coherent visions and plans into effective, implementable policies when pressures emerge that demand that something be done.

Potential paths forward will be constrained by politics, and politics is the art of the possible. What is possible, however, will depend on the state of opinion, and opinions, as they take form, are shaped by conversation.

An agenda for action therefore begins with talk.

CHAPTER 18

Changing Our Conversation About the Future

WHAT CAN ONE DO to bend the arc of history, and perhaps toward a better world?

In the transitions ahead, and as always in human affairs, actions will lead to intended and unintended consequences, some expected and others not. Actions, of course, are shaped by perceptions—not only perceptions of facts, choices, and potential outcomes, but also perceptions of interests, of how to judge the outcomes themselves. And yet a further step back, perceptions are shaped by shared opinions that emerge from conversations in homes, offices, and conferences; in journals, books, and editorials; in blogs and online discussions.

What we do and don't say matters in every sphere of life, and by our choices of what to say, all of us can help to shape the conversation that shapes the perceptions that shape the actions that in the end shape the world.

After years of seeing patterns of discussion repeat, I'd like to offer some thoughts regarding conversations, perceptions, and actions, and how one might nudge the world toward better outcomes.

THE PROSPECTIVE EMERGENCE of radical abundance will call for radical changes in expectations, perceived national interests, and plans—but not just yet. Today, the challenge of radical abundance is to develop and share a better understanding of the prospects ahead, to promote a more reality-based discussion of how to respond. That's why it's crucial to frame questions correctly and to focus first on questions that have clear answers.

Productive discussion on any topic requires keeping track of what the topic is. If no shared conceptual frame fits the topic, discussion is apt to be diverted by confused questions and misunderstood answers; if this confusion isn't challenged (which is to say, gently corrected), conversations can easily run off the rails and drag understanding down into a conceptual trap. This happened in the late 1990s and the debris from the train wreck is still being cleared.

In part because of this history, even the concept of what APM *is* can't be taken for granted just yet.* Through its links to the old nanotechnology mythos, the concept of APM is vaguely associated with a range of peculiar ideas. An essential point therefore bears repeating: Where the physical nature of APM technologies is concerned, the relevant questions pertain to the physics and engineering of compact, highly capable factories—not vague dreams, exotic products, or nanobugs, but a prospective manufacturing technology with implications for the human future.

Human action begins with discussion and a shared framework for thought. Getting the framework wrong can be fatal; getting the framework more-or-less right is at least half of what needs to be done. Thus, the first step in an agenda for action is talk that helps keep fundamentals in focus. Several key topics follow, but first a remark about enthusiasm:

---

* This is true not only for APM as a whole, but also for details that matter in science; these details reach all the way down to molecular reaction dynamics, where misunderstood scientific questions can draw attention away from engineering answers.

About Enthusiasm . . .

There's something that I feel I must say to some of my readers, and I hope that they will understand a somewhat counterintuitive message and take it to heart. If you find these ideas about prospective technologies compelling, convincing, and exciting—if you imagine vistas far beyond any I've outlined, or see solutions to urgent global problems and feel an urge to share the full measure of your excitement—then please lie down until the urge passes. In the world as it is, this kind of excitement triggers a negative response, and for reasons that usually make sense; almost all grand ideas proclaimed by excited proponents turn out to be wrong and are generally discounted without consideration. If you want to make a positive difference, please help to keep fundamentals first, help to correct mistaken ideas, and join the conversation without shouting.

And for other readers, please help to keep enthusiasts grounded and remind everyone else that feverish, misinformed people must not be allowed to set the agenda by provoking a backlash, or by fostering a kind of guilt by association. To allow this would amount to granting such people the power to control the agenda, but with a minus sign. This has happened before and must not happen again.

APM Is a Kind of Manufacturing

As a technology—as a basic, physical prospect—APM itself stands apart from its implications, just as silicon-chip fabrication stands apart from controversies about drones or Internet copyright law.

In the 1990s, the basic facts about APM-based production got buried in talk about wonderful spaceships, artery-clearing micro-subs, and robotic nanobugs with seemingly magical powers, all somehow confused with materials science.

We can't afford to replay this story. When questions revolve around APM itself—as a production technology—hypothetical applications, whether realistic or not, become a distraction. When a conversation starts to blur this distinction, one can help move understanding forward simply by drawing a line between APM itself and how it might be used.

APM Systems Are Factories, Yet Quite Different from Those of Today

In the past, when the physical nature of APM was forgotten, misleading ideas and techno-mythologies have slid in to fill the gap. The concrete picture needs to be kept in focus: An APM system is a factory in a box, a compact device packed with motors, gears, conveyor belts, and specialized gadgets of various sizes, typically plugged into an electrical outlet perhaps linked to a touch-screen interface. In other words, something much like a printer.

Inside the box, small, simple machines bring molecules together to build nanoscale parts, then larger machines bring small parts together to build larger parts, and at the end, machines of ordinary size bring parts together to build final products. The smallest machines guide molecular motion and bonding, while machines in succeeding steps do jobs like those of machines found in factories today. No magic, just large, complex systems built of enormous numbers of simpler machines.

This concrete picture can anchor discussion in reality at the point where discussion in the 1990s spun off into fantasy. To achieve this is simply a matter of answering the implicit question, "What is this?" by noting that an APM system is, in essence, a compact factory in a box.

Scaled-Down Machinery Enables Remarkable Results

High productivity and atomic precision are APM's unconventional features, and they stem from the characteristics of molecules and small machines. The connections are direct and easy to draw.

Small machines can perform many operations per second simply because small-scale machines can perform small-scale motions in proportionally shorter times. High-frequency operations, in turn, enable nanoscale machines to process many times their own mass per second, enabling extraordinarily high productivity in making atomically precise building blocks.

As for atomic precision itself, small molecules in raw materials consist of precise arrangements of atoms. Arrays of small, simple machines can bring molecules together precisely, joining them to make larger atomically precise components. Atomically precise components, if precisely assembled, will fit together to make atomically precise wholes—the principle spans the scale from small molecules to large objects.

In a technical context, by the way, one must often explain how atomically precise manufacturing operations can tolerate inevitable thermal fluctuations. The answer is straightforward: Stiff machines impose elastic restraints along motion paths—which is to say, they resist stretching and bending—and these elastic restraints can constrain the amplitude of thermal fluctuations tightly enough to strongly suppress unintended encounters between reactive molecules. (As discussed in Appendix I, the mechanical, kinetic, and thermodynamic constraints on reliability all involve exponential Arrhenius relationships, and error rates can be driven to levels below those seen in digital devices.)

Atomic precision has been practiced for more than a century, yet it can be extended enormously. Putting reactive molecular components together with atomic precision allows the production of extremely strong materials, high-efficiency solar cells, ultra-miniaturized computers, and more.

APM Technologies Parallel Digital Technologies

The digital information revolution provides a surprisingly good model for the prospective revolution ahead. Examining the parallels can help to frame better questions. Like digital information systems, APM systems contain billions of small, simple devices that operate at high frequencies, directed by software programs to produce patterns of indivisible units, bits or atoms. As with bits packaged in bytes and combined to form data files (with audio data for sound, pixel data for images, etc.), APM will combine atoms packaged in molecules to make patterns that serve useful purposes. And like other systems based on digital information (sound systems, printers), APM-level production technology will enable a box on a desktop to produce a virtually infinite range of products drawn from a global digital library.

———

THESE POINTS about the nature of APM frame the topic very clearly. If a technology isn't based on compact factories that contain factory-style machinery—or doesn't build products with atomic precision—then it isn't APM. Other useful technologies may be related to APM, but by definition, they aren't the same. When conversations get blurry on such basic points, one can contribute a lot by helping to cut through the fog, and I find that the digital analogy can be clarifying.

There's Been Surprising Progress Toward APM

An understanding of how current technologies can fit together reveals a surprising picture of progress. Without that understanding, however, progress toward APM-level technologies can easily be overlooked.

There's been a mistaken idea that atomic precision is something exotic, awaiting a breakthrough of some unspecified kind. In reality, progress has gone surprisingly fast and far, with no barriers in sight. Today, researchers can design and produce million-atom structures in billion-unit lots, structures defined at the atomic level. The challenge today isn't a matter of achieving atomic precision, but expanding its scale and scope.

"Nanotechnology" Has Become a
Brand Name for a Different Kind of Research

The great promise of nanotechnology is atomically precise manufacturing, and the US Congress established a program directed toward this objective, but the program instead did something entirely different. The program's leaders redefined "nanotechnology" and supported only nanoscale materials and devices, technologies as different from APM as cloth, cement, and wires are from a programmable digital computer.

Most research advertised as "nanotechnology" has therefore been irrelevant to what had been widely expected, and while atomically precise fabrication has flourished in the molecular sciences, people looking for progress toward APM-level technologies have been led to look in the wrong direction.

Near-Term Advances Won't Resemble Advanced APM

Despite enormous progress, there's still a long way to go, and in judging the distance, it is important to understand the landscape along the path. Indeed, one reason for misperceptions regarding the extent of progress toward APM-level technologies has been that steps on the path don't look much like the destination.

APM-level technologies will require precursors that themselves will require precursors—a chain of atomically precise fabrication technologies, each helping to build the tools of the next. The most direct paths lead through technologies that scarcely resemble APM, because they will at first be characterized by softer materials, smaller and simpler machines, different operations, products, and applications. APM-level results—large, low-cost, high-performance products—will require APM-level capabilities, which won't exist until near the threshold of technologies suitable for scale-up.

There is good reason to think that the pace of progress will seem modest at first, then will accelerate late in the game and become startlingly fast. It will be important to recognize where we are and what to expect.

APM Relies on Well-Known Physical and Engineering Principles

It's natural to assume that the radical implications of APM-level technologies must rest on some new principle, that they must, for some unspecified reason, embody a breakthrough in science. This mistaken assumption fosters a passive, wait-and-see stance regarding what APM can deliver.

To ask the right questions, it's important to keep the state of knowledge in focus. Current understanding of advanced APM rests on exploratory engineering and has an amply-documented physical basis.

Conservative exploratory engineering methods demand that every aspect of a design rely only on familiar principles in well-understood physical systems and—critically—that every engineering requirement can be satisfied with large enough margins to accommodate residual quantitative uncertainties.

It's legitimate to ask whether the functional roles in an APM system architecture can be filled in this way. Provided one asks the right questions—the physics-based questions regarding well-chosen, conservative options—I am confident that the answers are clear and positive. And in any technical review of these concepts, one must keep in mind the following proposition: *The key technical questions already have answers that can, in fact, be reviewed.*

It should go without saying that proposing an unworkable approach (and then rejecting it and saying nothing more) is worse than useless when a good solution is already known. To ask how an obvious problem has been solved (or avoided) is a natural part of inquiry—as is a search for unrecognized problems, for defects in the proposed solutions, or for better solutions than those yet proposed—but at this late date, in this field, to simply assume that an obvious concern has been overlooked is foolish.

Advancing Technologies Will Lead to Different Research Challenges

Different stages of development will raise different questions of science and engineering, and understanding these differences is a key to assessing research directions and judging the likely results. Here are some of the ways in which advances in atomically precise fabrication will shift the focus of research:

- A growing range of practical molecular building blocks will expand the scope of design to include finer-grained, more densely bonded structures.
- The key developments in design and modeling for molecular engineering will shift from biomolecular materials toward these more tractable structures.

- Guided molecular motion will gradually circumvent the problem of blocking unwanted chemical reactions, expanding the range of feasible operations in synthesis.
- Effective progress will increasingly require an engineering approach, and with this, new institutional frameworks and sources of funding.

Timelines, Pathways, and Applications Can't Be Fully Predicted

Estimated timelines for progress are subject to wide, legitimate disagreement. Timelines that involve human actions can't be predicted with confidence, and collective action can veer in unexpected directions (as is shown by the history of nanotechnology itself).

Likewise, potential paths in technology development can't be foreseen in detail, hence both difficulties and shortcuts may come as surprises (the recent leap forward in structural DNA nanotechnology, for example, was beyond anyone's expectations). Further, history shows how easy it is to overlook applications—witness what used to be roomfuls of computer capacity finding their way, and fitting, into pocket devices like cell phones.

In these areas, it would be unwise to adopt a pretense of knowledge. Timelines, pathways, and ultimate potential will remain persistent unknowns.

WHERE THE GROUND GROWS MORE CHALLENGING

The points above can go a long way toward grounding discussion and in themselves are reasonably straightforward, doing little more than helping to define APM and describing the nature of development paths. The ground becomes more challenging where physics and engineering meet global concerns. APM-level technologies promise to deliver an indigestibly large mass of capabilities. To frame a productive discussion, these capabilities must be regarded as linked together, and closely, because they share a common source.

With a focus on today's global problems, a compact outline looks something like this:

APM will radically expand the range and performance of potential products while radically lowering their costs. In addition, high productivity and atomically precise control of products and processes will enable deep reductions in waste and resource consumption—in part by dropping the costs of solar-based power and fuels below the costs of coal and oil—and will thereby provide a basis for sustainable material abundance on a global scale.

Even this brief outline of first-order consequences at a material level raises questions that go beyond what anyone can expect to answer or fully digest. The potential prospects include a falling demand for conventional labor, resources, and capital in physical production, with the potential for cascading disruptive effects throughout the global economy. Just one facet—the prospective ease of supplanting fossil fuels—is in itself a force great enough to change the structure of trade, geopolitics, security, and resource-driven conflict.

Outcomes will depend on technological change, but also, perhaps even more so, on expectations, perceived interests, and policies as they take shape in the years to come. Gradually, the prospective APM transition will rise from the level of an idea worth considering, to a prospect that demands contingency planning, to an emerging reality that calls for action at the highest levels of human attention.

In the preceding chapters, the reader has seen how potential developments connect with personal concerns, whether these involve studies, careers, or leadership responsibilities. There's no need to note all the ways in which prospects linked to the APM revolution may connect with interests in research, education, business decisions, grant proposals, and advising students, along with choices of what to read, whether to comment, and what to say. In each of these areas, however, personal concerns and spheres of action also connect with the flow of history.

EFFECTIVE ACTIONS, LARGE AND SMALL

Great tasks may seem to demand bold but impractical actions—actions that will naturally be set aside or postponed from day to day into the indefinite future, rather than being planned for this evening or next week. Great tasks, however, amplify the importance of smaller, more practical steps, and seemingly small steps have effects that may ripple, unseen, to the ends of the Earth.

Today, the most needed steps are those that nudge conversations toward considering an unconventional future while keeping discussion aligned with reality. Tomorrow, similar needs for nudges will arise, whether to broach further topics or to follow through on decisions, and at successively higher levels of intellectual and institutional engagement.

Practical actions span a wide spectrum. For example, promptly welcoming a remark made by someone else can be an act with power, because the first response in a group deliberation often sets the direction for the next. Even body language has power. A thoughtful look or a nod when someone raises a topic for consideration, or a skeptical look when someone speaks against it—these may scarcely be remembered (and pose little professional risk) yet can shift the scope of a group's conversations. Every new direction begins with small actions that open a door far enough to give a glimpse of what might be found beyond it.

Proposing or speaking out in support of an idea can of course do even more. In one context, a key step might be to float the idea that someone (somewhere, sometime), really ought to look into a question, or merely to suggest *cognizant* omission of a topic from a report because it is, for the moment, out of scope. In yet another context, the appropriate step might be to support a proposal to change evaluation criteria for research proposals, or to redirect a project, or to launch an initiative in national policy.

At the high end of personal investment and risk are commitments that direct the course of a life, whether choosing fields of study, committing to a career direction (in research, business, advocacy, and so on), or taking on a leadership role that mobilizes life-scale investments of effort by a few other people, a thousand, or millions. What counts as a practical, effective action depends on time, place, and personal role. And as always, prospects for enormous change don't overshadow individual action: They multiply its value.

LOOKING BACK AND LOOKING FORWARD

Unchanging physical law determines the form of a landscape more enduring than the seas and mountains of the Earth, a timeless landscape that defines the potential of physical technologies, and with this landscape, physical law defines an aspect of the potential of the human future. Humankind began its journey across this landscape many ages ago, a journey that has transformed human life.

From advances in biomolecular production (what we call farming) came settlement and civilization; from advances in machine-based production came industrial civilization; from advances in nanoscale digital circuits came today's Information Revolution.

Looking forward, we can see a molecular, mechanical, nanoscale technology that promises to change the material world as thoroughly as digital technologies have changed the world of information, and on a scale comparable to the Agricultural and Industrial Revolutions. Against this backdrop of historic change, we can see at least some of the implications of the revolution to come: potential solutions to world problems on the scale of global development, climate change, and resource scarcity—a potentially catastrophic success bringing disruptive change driven forward by compelling humanitarian benefits and (perhaps with more force) by compelling competitive advantages.

Among these competitive advantages are radical new military capabilities that threaten to drive an unstable and unpredictable arms race, and to do so at the very moment when objective national interests swing into closer alignment. This situation, unique in human history, calls for a reassessment of national interests and a creative exploration of strategies for cooperative, multilateral management of a revolution in economic, societal, and military affairs. Business as usual would bring needless, incalculable risks.

These opportunities and challenges emerge from applications of atomically precise manufacturing and its products, prospects that can be seen, at a distance, across the timeless landscape of technological potential. The landscape itself ensures that there are facts of the matter (some ascertainable, others not), while exploratory engineering provides a methodology for answering at least a few well-chosen questions about those facts—questions that, by design, can be answered based on system-level engineering analysis and conservative calculations based on textbook-quality scientific knowledge.

Meanwhile, through a discussion that embraces the conceptual foundations, cultural contrasts, and institutional structures of engineering and science, and examines a particular history of popular culture, scientific specialization, and national-level US politics, we've seen how these important and ascertainable facts about atomically precise manufacturing could be—and were—buried in misunderstanding and confusion for a decade or more.

This background and history highlights the crucial role of progress in atomically precise fabrication—ongoing progress, driven primarily by research in the molecular sciences—and also explains how that crucial progress came to be excluded from the scope of a redefined nanotechnology, placing us in a situation where enormous yet little-recognized capabilities stand ready to support surprisingly swift progress (along lines discussed in Appendix II). What has been missing is a focus that history has temporarily blurred.

An initial survey of productive technologies framed and grounded a picture of APM and its physical foundations, placing it among the other ways we make things today (exploring both similarities and radical differences), and presenting an intuitive yet quantitatively correct view of the world of APM-level nanomachines—how they would look and feel if magnified by a factor of ten million in both space and time.

Before this, I dipped deeper into the history of ideas (more to frame this book than anything else) by outlining how I came to have the concerns and sense of mission that drove me to explore the potential of technology, a sense of mission that today has led me to write these words. That path began with a high school student's peculiar concern about threats to the sustainability of our industrial civilization, and with the question of whether the future could be different from what had been expected. And industry itself, whether sustainable or not, exists to process matter to make things for people to use, and that matter consists of patterns of atoms.

═══

WHAT IF WE WERE really good at making material things? Our relationship with the material world would change in ways beyond imagining, yet in some ways familiar. We've seen the world of information transformed in a similar way, and in the span of a single generation. What has been scarce has become abundant.

Today, a radical abundance of symphony and song—and words, and images, and more—has brought luxuries that once had required the wealth of a king to the ears and eyes of ordinary people in billions of households. It seems that our future holds a comparable technology-driven transformation, enabled by nanoscale devices, but this time with atoms in place of bits. The revolution that follows can bring a radical abundance beyond the dreams of any king, a post-industrial material abundance that reaches the ends of the Earth and lightens its burden.

═══

WILL THE OUTCOME of the APM revolution in fact look anything like this? Understanding, confusion, choice, time, and chance will all have their say, and with a measure of wisdom, what follows may, in some measure, reflect our intentions. Outcomes aren't carved in stone, in part because human vision and choices are beyond prediction. As perceptions change, possibilities and politics change with them, while new media are transforming the discourse that shapes those perceptions. Unexpected new visions have again and again led societies in unexpected directions, toward surprising opportunities, problems, and sometimes solutions. As we sweep forward into the quickening rapids of history, will we find a safe course to a world worthy of the hopes of past generations? Neither success nor failure can be assured, but I am persuaded that our choices will matter.

And the first choice that matters is what to say, and to whom, today. What is your plan?

THE MOLECULAR-LEVEL PHYSICAL PRINCIPLES
OF ATOMICALLY PRECISE MANUFACTURING

THIS APPENDIX OUTLINES a set of concepts and principles that are central to understanding mechanically guided, atomically precise fabrication in a technical context. It is intended to help technically inclined readers mesh potentially unfamiliar concepts with their existing knowledge of scientific disciplines and physical principles.

I urge readers with backgrounds in the molecular sciences to recall or review the molecular-dynamics-based description of nanomechanical systems outlined in Chapter 5, which describes the basis for applying classical mechanical scaling laws across a range of nanoscale mechanical devices and compares and contrasts the behavior of a class of systems based on stiff devices with systems of the kind familiar in solution-phase chemistry and biomolecular machinery. Despite identical physics, the contrasts are sharp, and although professional-level knowledge of the molecular sciences is essential to a deep understanding of APM, especially where binding interactions and chemical reactivity are concerned, experience nevertheless shows that a casual, intuitive application of that professional knowledge can be profoundly misleading. Like many others, this topic has prerequisites and a range of common yet easily avoided intellectual pitfalls for the unwary.

The following sections describe the requirements for reliable mechanically guided chemical synthesis. Appendix II then describes a spectrum of potential physical instantiations of mechanically guided, atomically precise fabrication technologies, outlining a gradient of intermediate technologies between current laboratory techniques and advanced APM.

Atomic Precision Through Stereotactic Synthesis

Conventional, solution-phase chemical reactions are enabled by local structural features of molecules (which cause reactivity) and are directed to yield more or less specific results by those same features through their differential reactivities at potential alternative reaction sites. As synthetic targets become more complex, however, it can become difficult or impossible to direct reactions with sufficient specificity. Aside from intramolecular (and analogous supramolecular) reactions, solution-phase chemistry provides no control based on relative position per se.

Stereotactic chemical reactions can be regarded as intramolecular or supramolecular reactions in that linking structures direct reactions by constraining encounters among potentially reactive groups, separating some pairs while increasing encounter rates between others. Stereotactic methods differ in that the geometries of the linking structures are subject to more general, dynamic, discretionary control, yet at the reaction site itself these processes are essentially the same. The differences are nonlocal. For the specific structure provided by a shared covalent or supramolecular structure, substitute a mobile, mechanically articulated framework.

Stereotactic synthesis has been demonstrated on flat, crystalline surfaces, for example, through the use of STM tips to juxtapose specific reactive molecules and to remove specific hydrogen atoms to create reactive sites on hydrogen-passivated silicon (111) surfaces. Taking a broader view, one could perhaps argue that stereotactic synthesis dates from the emergence of ribosomes.

The Physical Requirements for Stereotactic Synthesis

As in stereotactic surgery, stereotactic synthesis effects structural changes at sites selected by means of positional control of an active component within a common frame of reference. Reliable stereotactic synthesis (of the sort most relevant to APM and precursor technologies) requires structures and operations that satisfy several potentially stringent design constraints:

1. *Structural stability:* Structures must resist degradation by thermal fluctuations (this design constraint requires sufficiently high unimolecular reaction barriers).
2. *Motion control:* Motions must avoid unintended reactive encounters (this design constraint requires both (a) appropriate machine kinematics and (b) sufficiently high mechanical stiffness, i.e., elastic restoring forces).
3. *Reaction rate:* Reactions must occur reliably (this design constraint requires both (a) sufficiently low reaction barriers and (b) sufficiently long reactive-encounter dwell times).
4. *Irreversible forward reaction:* Reactants must proceed to products irreversibly (this design constraint requires a sufficiently large negative free energy of reaction).
5. *Single reaction path:* Reactive encounters must lead to a single product (this design constraint requires sufficiently high transition-state barriers along reaction paths that would lead to unwanted products).

In other words, for reaction steps to succeed reliably, the reactive structures must (1) remain intact, (2) follow appropriate motion paths, (3) react within the time provided, (4) react only in the forward direction, and (5) yield a single result. All but constraint (2) are formally equivalent to constraints (or objectives) in conventional chemical synthesis, but the consequences of meeting constraint (2) greatly change the significance of the other constraints.

If these design constraints are each met by sufficiently large margins with respect to the relevant parameters, reactions can be highly reliable because failure rates decline exponentially as margins increase. Constraints (1), (2b), (3a), and (5) pertain to reaction rates that are governed by Arrhenius equations (where failure rates vary exponentially with the ratio of the reaction barrier height to the characteristic thermal energy kT); the barrier height in (2b) is determined by an elastic energy that increases in proportion to the mechanical stiffness and (crucially, for low-stiffness machines) with the square of the displacement tolerance. Constraint (4) pertains to thermodynamically-controlled product ratios, exponential in free energy, and given condition (4), the dwell time (3b) governs a kinetically controlled transition (in which failure rates decline exponentially with both time and barrier height).*

The remaining constraint (2a) simply requires that equilibrium positions along the nominal motion path approach the intended transition state while avoiding others by margins that (considering thermal fluctuations, elastic forces, and displacement tolerances) meet constraint (2b).

The degree of difficulty of meeting these quite general conditions depends on both the characteristics of the intended operations and the context-dependent, system-level reliability requirements. In some contexts (as discussed in Appendix II), these conditions are, in quantitative terms, stringent; in others, they are relaxed.

A Broad Spectrum of Potential Systems and Methods

It is important to keep in mind that stereotactic control of chemical reactions is a general and broadly applicable concept, not tied to particular classes of products, reactions, reaction environments, or kinds of machinery. The most challenging, high-performance objectives involve

---

* To get a sense of the magnitudes involved, note that with a strongly favorable free energy differential $\Delta G = 100$ kJ/mole (about one-fourth the energy of a typical carbon-carbon bond), the Boltzmann factor $\exp(-\Delta E/kT)$ is $\sim 10^{-18}$ at room temperature; note also that employing n conditional repeated trials can reduce required free energy differentials by $1/n$.

products that consist of densely bonded covalent solids, built via reactions involving high-energy reactive species, with operations performed in inert environments (e.g., vacuum) and directed by means of complex machines with components that themselves consist of densely bonded covalent solids; in this class of systems, r.m.s. thermal displacements < 0.02 nm are both desirable and achievable. The most accessible objectives, however, are those closest at hand, for example, structures that consist of polymeric building blocks, cross-linked via conventional reactions, with all stereotactic operations performed in aqueous environments and directed by means of relatively simple devices that consist primarily of self-assembled foldamers; this class of systems includes stereotactic devices that can accept r.m.s. thermal displacements > 2 nm (this relaxes stiffness constraints by a factor of ten thousand compared to the challenging case outlined above).

Pure self-assembly marks the endpoint of a spectrum of stereotactic methods in the limit where mechanical constraints go to zero. Pure self-assembly requires encoding the structure of each product in the structures of its parts, placing the entire burden of specificity on the design of the building-block components themselves; the first step into the spectrum of stereotactic methods lifts this burden by applying loose mechanical constraints to direct self-assembling (or self-aligning) components to particular regions of a structure. With this step, complex products no longer require complex components.

Large, self-aligning building blocks, loose positional tolerance margins, low stiffness materials, simple machines, and simple motion constraints constitute a coherent, accessible technology base that can enable further steps along a gradient of technologies that extends to the also-coherent but as-yet inaccessible technology base of advanced APM-level production: small building blocks, tight positional tolerance margins, high stiffness materials, complex machines, and complex motion constraints. Appendix II discusses this gradient and directions for further progress.

INCREMENTAL PATHS TO APM

INCREMENTAL DEVELOPMENT pathways lead from current technologies to APM; these are marked by advances in constructing AP mechanical systems and implementing next-generation systems for performing atomically precise fabrication. The key advances center on stereotactic assembly, an extension of the age-old method of building things by moving parts into position, but applied to assembling nanoscale parts with atomic precision. Appendix I discussed the general physical constraints that must be met at each stage; this section describes coherent sets of intermediate technologies that can meet those constraints while moving forward.

METRICS FOR PROGRESS, PAST AND PRESENT

Across a span of centuries, machine technology has climbed a gradient in performance, and along the way several kinds of metrics have been prominent. They include:

- Increasing quality of materials
- Increasing control of the shape of components
- Increasing complexity of feasible systems

The Industrial Revolution has required advances in precise, intricate, high-performance machine tools, and hence in the metrics above:

- Regarding materials, measures of quality include uniformity, strength, and stiffness; these led to the use of steel for building machine tools.
- Regarding control of shape, requirements for increasing precision led to precision machining of the steel components used to build high-precision machine tools.
- Regarding complexity, machines with many precise moving parts have enabled high-throughput automation.

Supported by this steel-centered production technology, technologies of every kind have grown and proliferated, yielding progressively better rubber tires, silicon chips, and the like—technologies that are themselves paced by progress in materials and the means of shaping them.

Potential advances in AP production technologies climb a similar gradient, and in nanoscale production machinery key metrics for improvement are much the same:

- Regarding materials, uniformity and stiffness are once again critical to building high-performance machine tools.
- Regarding control of shape, however, the challenge differs, centering not on precision, but on freedom to choose desirable forms.
- Regarding complexity, precise complex machines are once again essential to high-throughput automation.

These metrics are interrelated, and climbing the AP technology gradient will tend to move a range of technologies forward together.

EXPLORING PATHS UP THE AP TECHNOLOGY GRADIENT

Each level of technology relies on a technology base—a set of enabling technologies that determines what products can be made. A technology base can serve a range of applications (for example, machine technologies enable production of cars), but what is more important, a technology base must also enable production of its own technological components— the machines that make cars, the machines that make the machines that make cars, and so on. Fundamental progress comes from using a technology base to implement the components of an improved technology base, thereby advancing a step higher in a direction that enables further steps.

If small steps are sufficient, but large steps are possible, then a prospective path can be followed; the size of the actual steps taken will depend on how well vision, practicality, invention, and investment fit together. Regarding feasibility per se, if one finds a continuous gradient leading upward, then small steps will indeed be sufficient, and the usual criteria for robust exploratory engineering will be met, with the usual caveats that go with conservative assessments (in particular, larger steps will be likely in practice).

To understand the overall gradient of potential technologies between here and APM, it will be useful to consider several dimensions of capability and consider how they are linked. With the above-mentioned metrics in mind, and working from the bottom up in a systems-engineering sense, we can consider, first, levels of structural quality (materials and components), then the levels of control needed to produce these structures, and finally what can be built at the systems level, as measured by feasible complexity and performance.

A Gradient of Quality for Materials and Components

As outlined in Chapter 12, the most complex AP structures that can be engineered today consist of folded macromolecules. These include engineering-grade polymers such as polypeptides (the protein molecules of horn, silk, and many biomolecular machines) and some of their nonbiological relatives (peptoids and other foldamers). These molecules can provide intricately shaped components that are able to direct self-assembly among both similar and quite different molecules. Nucleic acids and their analogs provide materials for building another class of folded macromolecular structures, exemplified by structural DNA nanotechnology, which can construct site-addressable AP scaffolds on a million-atom scale.

Polymers used this way—folding to make AP objects that can serve as building blocks—enable construction by the methods at one end of the spectrum outlined in Appendix I: pure, unguided self-assembly, in which each member of a collection of building blocks can encounter all of the others, forming a specific structure if and only if the blocks are large and complex enough to bind together in only one way, like a jigsaw puzzle.

Working with today's accessible self-assembling structures limits control of shape and of related surface properties such as charge and polarity. They are assembled from relatively large building blocks (for example, tens of cubic nanometers or more), while their polymeric materials are built of units—monomers—that are considerably larger than an atom, and the requirements for folding further interfere with choosing the shapes best suited to particular functions. Finally, self-assembling structures of this kind have low stiffness; they are much like typical engineering polymers, and hence are sparsely bonded.

At the high end of the gradient one finds materials that include oxides (with local patterns of atoms and bonds like those of glass and ceramics) and covalent solids (with local patterns of atoms and bonds more like those of silicon, diamond, and graphene). As a consequence of their dense networks of bonds, these materials have high stiffness; they are fine grained, being built of monomers (often called "growth species") that contain no more than one or a few atoms or ions; and because they are fine grained, they enable extensive freedom of design—provided, of course, that means are available for directing the assembly of precise, complex, nanoscale structures, even though their building blocks are small, identical, and unselective in the ways that they bond.

Regarding the gradient itself, the key observation is that building blocks need not be at one extreme (sparsely-bonded, coarse, complex, macromolecular objects), nor at the other (tightly bonded, fine-grained, simple, near-atomic-scale monomers). They can be of any size between these extremes and can have intermediate properties according to all of these metrics.

In particular, toward the middle of the spectrum are building blocks with sizes in the range of tens of atoms: medium-grained (large enough to orient and bind selectively, though not complex enough to fit in a unique place in a structure); medium-stiff (in comparison to typical foldamers, more densely bonded inside and more densely bonded to neighbors); and enabling extensive design freedom (because they are finer grained and not constrained by the rigors of design for self-assembly).

Potential building blocks toward the middle of this range (a band in the gradient) include polycyclic, nanometer-scale covalent molecules with three or more functional groups that can form cross-links to neighbors. For present purposes, cross-linking includes not only typical covalent bonding, but also coordination to shared metal ions. Here, metal-organic frameworks provide suggestive examples; metal-binding proteins provide others.

As the above discussion suggests, improvements in stiffness, granularity, and design freedom are closely linked to improvements in materials quality and shape control. Together, these properties enable improvements in metrics that include performance and ease of design, and these, in turn, affect the feasible level of system complexity. It seems that every important metric places mid-size blocks in the middle.

A Gradient of Levels of Stereotactic Control

Conventional self-assembly requires no stereotactic control because blocks fit together in only one way; assembly of the simplest fine-grained blocks requires stringent control, because blocks may bind almost indiscriminately; assembly of mid-size blocks requires only loose stereotactic control to direct building blocks to a locale, because localized self-assembly (or self-alignment) can provide more specific targeting. Thus, the distance between alternative binding sites—the acceptable margin of error in positioning—can be of a magnitude that ranges from the diameter of the product itself (toward the pure self-assembly limit) to the diameter of a building block (where small blocks will lead to closely spaced reaction sites).* Here again we find a gradient, one that strongly affects the requirements for machines.

In stereotactic positioning systems, the frequency of thermally induced misplacement errors depends both on the acceptable error margin (the frequency falls exponentially with the square of the distance between alternative targets) and on the stiffness of the structure that resists misplacement. The scaling law is such that a ten-fold increase in the acceptable error margin reduces the required stiffness one hundred–fold.

Looking forward from the low end of the gradient, design exercises show that self-assembled devices made of relatively soft blocks of polymeric materials can be sufficiently stiff to enable reliable stereotactic assembly of smaller blocks of more rigid materials. Devices built of these stiffer materials, in turn, can provide more stringent motion control, enabling the assembly of yet stiffer materials.

Conversely, looking back from the high end of the gradient, design exercises show that the requirements for reliable stereotactic assembly of fine-grained, extremely stiff covalent structures can be met by machines that are substantially less stiff and consist of coarser-grained materials that can be made with less stringent stereotactic control.

Thus, where materials and stereotactic control are concerned, both ends of the spectrum are part of a path: From the high end a path leads back and down to more accessible systems, while from the low end a path leads forward and up. These paths meet in the middle, leading through blocks of intermediate size, stiffness, and bond density.

A Gradient in Environments and Means of Transport

Potential environments for stereotactic assembly range from water to organic solvents to the ultimate low-viscosity, chemically-inert environment—empty space with no fluids at all.

---

* Or even less: The distance between alternative binding targets can be less than an atomic diameter in densely bonded covalent solids (and can still be compatible with reliable targeting).

Fluids provide two related and useful services. Fluids give nanoscale objects the local mobility required for self-assembly and the long-range mobility that enables molecules to travel from a source to a destination by diffusion, without conveyors or pipes. In particular, at present, water has a special attraction because it's a well-explored medium for molecular folding and self-assembly.

There are advantages to full, end-to-end motion control (as in the advanced APM systems described in Chapter 10), yet these advantages either pertain to secondary considerations (such as speed and energy efficiency) or to pushing the boundaries of feasible chemistries (which may require, for example, non-contact conveyance of unusually chemically reactive species).

One can, however, achieve stereotactic control of assembly without what might seem to be a prerequisite—that is, without having to guide the motion of building blocks at all. The overriding objective is to control where blocks bind, not to control how they reach the intended location, and in fluids this opens a possibility not found in macroscale manufacturing: Let diffusion transport the blocks and use stereotactic control to activate or deprotect a chosen location. In this approach, a catalytic structure can serve as a tool to make stereotactically directed modifications to a target.

This approach should be familiar to chemists—it amounts to a strategy for using protecting groups to block potential reaction sites during a series of steps, then removing these protecting groups to activate just one or a few of these sites. This method works across the spectrum from coarse-grained blocks to atoms; the smallest building blocks can be transported as gases, while the protecting groups that block sites can be as simple as hydrogen atoms. Indeed, reactive gases together with hydrogen-based protection chemistry are the basic elements in the Zyvex method of building precisely modified silicon crystals, which uses an STM to prepare active sites by removing selected hydrogen atoms.

Even where AP machinery performs the final block selection and positioning, a fluid can enable long-range transport by means of diffusion, while the final movements can rely on thermal motion and self-alignment, again enabled by a fluid environment. By exploiting fluid transport this way, machine architectures can be relatively simple and can tolerate (and exploit) large-amplitude thermal motion.

Here again, working with stiff, fine-grained, covalent materials can require more stringent conditions. As presently understood, the required chemical operations will require inert (e.g., vacuum) environments with mechanical transport of highly reactive molecular species.

Other approaches, however, offer easier access to relatively stiff, fine-grained materials. Iron pyrite (FeS_2) and many silicates, for example, are about ten times stiffer than foldamer-based materials and, in addition, assemble spontaneously from aqueous solutions. Building structures by guiding the assembly of materials that form under mild, aqueous conditions offers advantages that include solution-phase transport of building blocks and compatibility with (for example) familiar classes of foldamers.

A Gradient of Complexity

The low-end systems of the sort just described can build components, but provide no stereotactic means for assembling components to build complex systems. However, improved stereotactic control enables improvements as measured by many metrics, and these include improvements that facilitate the design and fabrication of complementary surfaces. By expanding the range of feasible product structures, fine-grained stereotactic synthesis can enable advances in component-level self-assembly, thereby extending the applicability of the very methods that compete with component-level stereotactic assembly.

As before, a gradient links self-assembly to advanced stereotactic assembly, and what are initially small, simple machines can therefore support progress up a gradient of system size and complexity.

Improvements in component fabrication will enable expansion of the range and performance of functional devices. Examples include transducers (chemical, optical, electrical, etc.) that can drive controlled motions at higher frequencies and with a greater number of distinct input control channels, devices for materials handling that can bind and release selected building blocks from a mixture in solution, positioning mechanisms with stiffer structural components that enable more precise and reliable motions, and use of finer-grained, higher-valence building blocks.

The expanding range of feasible products will, of course, support a range of practical applications. This has been the story of advances in AP fabrication since its beginnings more than a century ago, and it will continue.

Small AP structures have wide-ranging applications. Potential early products include tailored, high-performance catalysts, specialized molecular production lines modeled on modular polyketide synthases, and programmable oligomer-synthesis mechanisms modeled on ribosomes. These small, specialized, high-throughput production machines are modeled on devices of known utility in areas that range from basic science to fine chemical and pharmaceutical synthesis. Small structures can also serve as components and building blocks for larger-scale systems in which nanoscale components are organized by self-assembly to form active structures for display screens, lighting, photovoltaic cells, and computer memory devices, among other potential examples.

Because chemical and biochemical methods can make products in substantial quantities, and early generation production machines can be similar, every step up the gradient of production technologies has the potential to deliver substantial results, as measured by mass. Today, analogous processes yield products on scales that range from micrograms to tons, and even a microgram can contain trillions of nanoscale objects.

The most efficient production systems are specialized, and this is true regardless of scale. A natural approach involves using smaller, specialized production machines to make diverse building blocks in quantity, then larger machines (perhaps less specialized) to make a yet wider range of products that consist of combinations of parts.

The picture that emerges is a network of production systems not unlike the network of specialized factories that deliver parts that other plants assemble to build a range of automobiles, or the network of specialized chemical plants that deliver chemical intermediates that other plants use in synthesizing diverse pharmaceuticals. Monomeric building blocks delivered for use in synthesizing foldamer chains; nanoscale building blocks delivered for loosely guided stereotactic assembly; microblocks delivered for high-throughput, end-product assembly by APM systems—all these follow the same basic pattern—specialized production of components to support more flexible production of end products—and they follow this pattern the same basic reasons.

Note that none of these systems requires machines that can make all the parts needed to make similar machines. Like today's global industrial technology base, an AP technology base as a whole may have this capacity, but its simpler component production systems do not. Among other advantages, this organization facilitates smooth, incremental upgrades as the aggregate technology base progresses along a gradient of capabilities.

LOOKING BACKWARD FROM APM

To work toward an objective, it often helps to work backward, asking what must come before it, and what must come before that. Chapter 10 outlined an architecture for large-scale, high-throughput APM systems, but systems of this kind, and even their components, are still well beyond reach. What kinds of systems could serve as precursors, and what would be required, in turn, to bring these precursors within reach?

One could build the components of large-scale APM systems by means of systems only half as large, and these smaller systems, in turn, could be built using systems half as large as those, and so on, down to a very small size. Where is the stopping point in this backward-looking conceptual process, the physical starting point, in a forward direction, for a scalable APM-level technology?

The best way to approach this question is to ask how much could be discarded while still retaining APM-level production capacity—not measured by scale, throughput, or efficiency, but instead measured by quality, by the kinds of components and systems that can be produced. Thus, one might ask, "What would a minimal core system require to make and assemble the kinds of components needed to make the smallest and simplest APM-*quality* production systems?"

One straightforward way to reduce a core system's size and complexity is to outsource or omit operations to minimize required functions. For example, production of high-quality molecular-scale building blocks can be outsourced to other parts of a contemporary production network, hence a core system can omit purification and most stages of chemical processing. Likewise, because a fluid medium can carry molecular-scale building blocks to sites near where they are used, complex transportation systems can be omitted. Similarly, slow and unspecialized machines can assemble blocks in patterns to make a wide range of components, hence specialized, efficient, high-throughput systems like those described in Chapter 10 are unnecessary. Finally, directing unspecialized machines to construct a range of components requires a series of signals—a sequence of instructions that serves as a program—but an external source can provide a sequence of instructions, hence stored data and digital control systems can be omitted.

This set of simplifications leads to an architecture for APM-quality production that is quite unlike that of an advanced APM system. With inputs prepared externally, transported by fluids, and put in place by machines guided by external programs, there's relatively little complexity in any device. Small building blocks, intermediate building blocks, and complex products would be made in different stages and places, in a network of production systems.

These simplifications sacrifice scale, throughput, cost, and efficiency, yet at the level of synthesis and assembly the operations performed can be much like those in a full APM system. Similar molecular transformations and assembly operations can build structures of similar quality. Thus, systems designed along these lines could be used to build the kinds of components and systems that define and enable scalable APM-level technologies. Systems at this level of capability would mark a threshold for the smooth scale-up process described above.*

To embody this level of technology, no individual machine system need be as large as a cubic micron; in aggregate, however, a network of production systems might be built on an industrial scale, using macroscopic quantities of nanoscale devices to deliver tons of nanoscale products useful in materials, electronics, medicine, and elsewhere.

---

* The threshold could no doubt be approached and crossed in other ways as well; this discussion is an exercise in system-level exploratory engineering, not in prediction, optimization, or program management.

Meeting in the Middle

To implement systems at the level outlined above still requires AP fabrication well beyond today's technologies, but the task doesn't require exotic materials or an ultra-clean environment. The required components (moving parts, actuators, binding sites . . .) don't require extraordinary performance and can be made of materials compatible with solution-phase synthesis (some of which are, in fact, extraordinary). For example, AP graphene nanostructures have been produced through solution-phase chemical synthesis, and a range of stiff, fine-grained materials can be synthesized in mild, aqueous environments. (Materials of this sort are often characterized by mixed ionic/covalent bonding.) Systems built of these relatively accessible materials can then be used to make components that require more stringent conditions for chemical stability and motion control. Working backward:

- Components made of the stiffest, covalent materials can be produced by machines built of more accessible materials.
- Components made of these more accessible materials can be produced by machines that work in a solvent.
- Machines that work in a solvent can build machines that work in dry environments.
- Machines made of stiff components can be assembled through soft stereotaxis aided by self-alignment.

In short, production systems at the threshold of advanced APM-level technologies can be built by production systems based on technologies several steps back from that threshold.

Thus, paths can be seen that lead backward from advanced APM and forward from self-assembly. These paths meet in the middle. Along the way there's no single, extraordinary machine, nor any particular, critical technology. Instead one finds prospects for a rising AP production technology base, a growing range of components of increasing performance, and a progression of systems for assembling AP parts into larger and more complex products.

Further, nothing about the technologies involved necessarily limits the scale of production at any stage of development. Instead, one finds a technology base that extends ongoing advances in chemical and bio-chemical synthesis—processes that have often been scaled up to make milligrams, grams, or tons of AP products. Because each stage can enable large-scale production of tools for next-round production technologies, these next-round technologies can likewise enable large-scale production.

This discussion has focused on using tools to build better tools, with less emphasis on direct applications to products along the way. If history is any guide, scalable production of new classes of AP products will find applications beyond those we can imagine today.

With smooth paths forward, a compelling objective, increasing re-wards along the way, and a world in which many centers of technological competence pursue many goals, I am persuaded that we will, in fact, see ongoing progress along the general lines outlined above. Further, there are good reasons to expect the pace to quicken.

A QUICKENING SPIRAL OF PROGRESS

What sets the pace of progress? Investment and creativity are crucial factors, but so is the nature of the technology being developed. On paths toward advanced APM, the nature and pace of the development process will change as much as the technologies themselves.

Faster Production and Testing

The pace of development depends in part on the time required to make and test a new product, and by the standards of large engineering projects both the cost and the time required to make a new macromolecular structure are typically modest, not millions or billions of dollars spent over times measured in years, but thousands of dollars spent over times measured in days or months. Advancing production technologies, to-gether with standardized building blocks, can be expected to shorten the time required to go from a design to a working artifact. For production systems that execute operations at rates in the range of hundreds to millions of cycles per second (the range from biomolecular to advanced-level systems), an hour is a long time.

The same considerations apply to testing, because the rate of molecular processes can frequently enable fast tests of product performance. (For a comparison to a complex, macroscale engineering project, consider how many months may be needed for individual tests of a rocket, and then the cost of revising a design and running another trial.)

Faster Design

Improvements in the design process play a triple role in setting the pace of progress. A reduction in design time reduces the delay between concept and execution; greater design predictability reduces the number of design-and-test cycles required for success; and finally, the level of design capabilities helps determine how large each development step can be—if large departures from previous products can be designed with confidence, fewer steps will be needed to reach an ambitious objective.

Critically, advances in fabrication can make results more predictable, and predictability increases the power of design. Today, despite advances, the design and fabrication of foldamer-based molecular objects remains difficult. The advent of effective stereotactic methods, however, will lead to a qualitative change in design, fabrication, and predictability. Using standardized building blocks in designing finer-grained, more regular structures simplify design tasks; using direct positional control to assemble complex patterns of parts largely decouples design from production; and building with stiffer, more stable materials makes results more predictable.

Computational tools will offer progressively better support for engineering with wider and more accessible ranges of materials. In particular, computer-aided design tools patterned on current software used by mechanical engineers will make many aspects of nanoscale mechanical systems design as tractable as designing conventional mechanical systems is today. Considering the domain of fine-grained structures built of covalent solids, I can say with confidence based on personal experience that designing components and devices suitable for APM-level mechanical systems is a straightforward task (even with standard computational modeling software), and also easy to teach.

EXPLORATORY OPEN-SOURCE DESIGN

It is safe to predict that some design tools—and perhaps many designs—will be open-source, and that some of these will be applied to the design of systems that aren't yet within reach of physical implementation. Indeed, until the financial crisis hit in 2007 I had been working with Nanorex, a company developing an open-source application for AP engineering, NanoEngineer-1. The Alpha version, developed in conjunction with experimentalists, offered prototype capabilities for structural DNA engineering and got good reviews. NanoEngineer-1 also provided design tools for mechanical devices based on covalent materials, interfaced with standard molecular mechanics and molecular dynamics code (which for reasons discussed in Chapter 5 can be remarkably effective for testing how such devices behave). Similar software development projects will surely be pursued in the future, supporting both contemporaneous experimental research and forward-looking engineering exploration.

Open-source design software can be expected to support crowd-sourced design organized as competitive games. The design work involved can be fun (based on my experience and what I've seen others do), and the results yield attractive, visual products, both pictures and videos. Further, designs can be evaluated by quantitative engineering criteria and metrics. Competitions can use these metrics as scores.*

In short, we will likely see both open-source design tools and open-source design development, and the results are likely to be weighted toward high-end technologies. These design tools and results will accelerate progress when fabrication tools become able to implement designs already developed.

---

* The Baker group at the University of Washington has packaged computer-aided protein fold prediction as a game, Foldit, which has been a remarkable success as measured not only by participation (many players, no scientific background required), but also by results and scientific papers. Foldit makes protein folding fun—it's a kind of puzzle-solving problem—and proposed solutions earn scores in an international competition.

No one can predict the extent to which open-source tools and designs will accelerate advanced development; this is yet another contributor to irreducible uncertainties regarding the pace of progress.

On Estimating the Pace of Progress

One might seek a conservative estimate of the time required to develop APM-level technologies. There is, however, no single notion of what constitutes a conservative estimate. When considering potential benefits, a conservative estimate will emphasize potential delays in implementing programs and achieving results. When considering potential risks, a conservative estimate will emphasize the potential for fast progress, perhaps in unknown locations that are hard to monitor, or perhaps in an urgent, international program launched in part to forestall catastrophic climate disruption.

This double-sided notion of conservatism must be kept in mind when considering how uncertainties add up. And at a more reflective level, one would be well advised to attempt to correct for well-documented tendencies toward overconfidence in making judgments, resistance to accepting uncertainty, and inertia in revising assumptions. Without attempting to attach years or dates to the discussion, it may be useful to review some of the considerations.

Delays in Research Decisions, Planning, and Coordination

At the moment, for reasons both of scientific culture and history (discussed in Chapters 8 and 13), advances in AP molecular fabrication (discussed in Chapter 12) have been under-exploited for many years, and now far outstrip their potential applications in AP systems engineering. The reasons for this lag will not vanish in an instant.

Developing effective research programs requires time and effort, both intellectual and organizational. Translating opportunities into concrete plans requires inspiration, evaluation, discussion, and a measure of agreement before signing purchase orders for equipment, hiring research staff, structuring new collaborations, or developing programs.

Progress will depend on the state of opinion in communities that may not yet have coalesced. Shared vision and drive can telescope stages; misunderstood objectives and weak motivation can delay progress indefinitely.

In this connection, thinking in terms of averages or typical cases can be misleading because there is an asymmetry between forces that drive progress and forces that cause delay. What sets the pace is action, not inaction; not the general state of opinion or knowledge, but the emergence of a critical mass of resources—competence, coordination, and funding—wherever this critical mass may arise. With billions of dollars now flowing into nanotechnology programs worldwide, but directed toward research unrelated to the promised APM vision,* it would not be surprising to find that a better understanding of both means and ends mobilizes substantial resources to support research programs that are structured to deliver results.

Unexpected Difficulties (and Advances) in Laboratory Research

As any practitioner or attentive observer of laboratory research knows, progress in research can be painfully slow relative to naïve expectations, and work often meets unexpected difficulties that require a new approach or a change of objective.

Here again, however, there's an asymmetry to consider, one that makes individual laboratory experience a potentially misleading guide to the pace of progress in the world as a whole, or in well-funded, well-managed research programs. When multiple groups pursue different approaches to satisfying a set of system-level engineering requirements, different groups will encounter different difficulties and it becomes less likely that all will encounter an intractable problem. (In other words, redundancy increases reliability.) Conversely, when multiple groups pursue different approaches, the likelihood increases that one will find a solution that is better or earlier than a single research group could expect. Thus, in a systems-engineering context, expectations calibrated

---

* Recall that Chapter 13 documents the disconnect between the promises made to Congress (advanced atomically precise technologies) and the quite different direction of the subsequent US national nanotechnology program.

by laboratory-level research experience will tend to be miscalibrated, overestimating the likelihood of encountering roadblocks and underestimating the likelihood of discovering shortcuts.*

It is important to keep in mind that every functional requirement between here and APM-level technology has many potential implementations that, when considered in context, make no stringent physical demands. The essential challenges are overwhelmingly those of design, with the greatest challenges today centering on the complexities of macromolecular folding and self-assembly. Further up the technology gradient, these challenges decline.

The Downstream Challenges of Systems and Complexity

When questions center on downstream developments, it becomes important to step back from experience in the molecular sciences and to turn attention instead to experience in conventional forms of engineering, the practice of developing complex systems based on systematic assembly of well-understood components in accord with the principles of modular design.

Advances along the technology gradient will enable increasingly straightforward, predictable design using conventional engineering methodologies. Along the way, the challenges of systems design will increase together with system complexity, but only as fast as the enabling technologies permit. From today's laboratory perspective, where small gains are often hard-won, assembling numerous parts with predictable results may still be hard to imagine. Sound judgment requires care in matching prospective challenges with their appropriate technological contexts.

Stereotactic control of chemical synthesis will change the game from the bottom up, enabling greater freedom of design, facilitating self-assembly, expanding the range of accessible materials, expanding device performance and functions, and enabling improved stereotactic methods. Open

---

* Note that Moore's Law progress in semiconductor fabrication has continued, successfully, for this very reason. Each new generation of fabrication technologies has a set of abstract requirements—more precise wafer steppers, shorter-wavelength light sources, compatible optics, masks, photoresists, and all the rest—and at every stage, multiple groups (first researchers, then corporate development labs) have pursued different ways to satisfy each requirement.

paths lead from current laboratory capabilities through simple devices and then, by stages, to advanced APM. There is no gap.

IN CLOSING

The pace of progress toward APM capabilities will depend on both the inevitable delays in implementing and pursuing research programs and the acceleration that can follow from many teams working in parallel to solve inherently tractable problems. At every turn, the pace will be set, not by average, typical, or worst-case results, but by the shortest paths found—if solutions to problems are shared.

In a world with many cultures, nations, and decision makers, there are many potential centers of competence capable of advancing development, regardless of how many others may stand aside. Thus, APM-level technologies may seem hard to achieve, yet in another sense they seem hard to avoid.

Joining collaborative development efforts can accelerate progress, bringing forward solutions to problems in health care, environmental protection, and global development. Among these are capabilities sufficient to restore the pre-industrial composition of Earth's atmosphere and reverse the drivers of climate change.

Further, as we've seen in discussing potential disruptions, collaborative development may be necessary in order to avoid needless risks of conflict. Carefully structured collaborative development efforts can avoid spiraling uncertainties while providing a basis for coordinated policies that can serve shared, compatible, and better-aligned national interests.

Depending on whether the question is one of risks or of opportunities, conservative estimates of the pace of progress may be relatively fast or relatively slow; neither assumption can be taken for granted. Although physical considerations place prospects for APM on firm ground and provide a general view of the paths from here to there, human considerations—creativity, chance, and choice—permeate other aspects of the prospects ahead. Alternative research paths, the likely pace of progress, and how best to manage development and the prospective transition—all these and more are questions to weigh and discuss.

ACKNOWLEDGMENTS

A BOOK OF THIS SORT must draw on the resources of many minds, and because it builds from molecules and machines to economic, environmental, and international affairs, those minds and their contributions must be diverse.

The Oxford Martin School, a unique, interdisciplinary research community within Oxford University, has helped to provide that diversity. The school's programs address twenty-first-century challenges and opportunities that range from economic development to nanomedicine, food, human rights, and the governance of geoengineering; I thank its director, Ian Goldin, for welcoming me for an extended stay, and Nick Bostrom for opening the door by inviting me to join the community at the Programme on the Impacts of Future Technology. Within that programme, I thank Nick along with Anders Sandberg, Toby Ord, Stuart Armstrong, Daniel Dewey, Seán Ó hÉigeartaigh, and others for more than a year's worth of wide-ranging and challenging discussions. During this time, it has also been a valuable pleasure to share space, tea, and ideas with members of the Oxford Uehiro Centre for Practical Ethics.

313

Reaching back a few years, I thank Dennis Pamlin, then Global Policy Advisor for the World Wide Fund for Nature, for initiating and supporting a study of the implications of atomically precise manufacturing for the environment and global development; developing the report with his help did much to clarify the issues addressed in this book. I thank Alex Kawczak of the Battelle Memorial Institute for launching and inviting me to serve as co-leader of a roadmap development project that explored paths from today's atomically precise fabrication technologies to the threshold of atomically precise manufacturing, and I also thank the roadmap's many contributors from academic and corporate research groups, and from the hosting U.S. National Laboratories. During these years, George Church, Paul Rothemund, Tom Theis, and Ron Zuckerman (among many others) have provided stimulating discussions of prospects and current research in atomically precise fabrication.

Longer term (across many, many years), I have learned from audiences at university lectures and seminars, and from participants in scientific, engineering, corporate, military, and environmental meetings on several continents; these extraordinarily diverse audiences have called for approaching a multitude of topics from a range perspectives, and their questions and comments have helped to shape the concepts presented in this book. On a par with these discussions, and almost as wide-ranging, have been conversations with Mark S. Miller in walks through the hills overlooking Silicon Valley.

Finally and most directly, I'd like to thank those who helped to guide this book forward and bring it to publication: my editor at PublicAffairs, Brandon Proia, and literary agent, Loretta Barrett, who both understood what I was trying to do, and Robert Silverman and Stephanie Corchnoy for smoothing the content and the text. Most of all I would like to thank my wife, Rosa Wang, for helping to shape this book in its scope, purpose and content, and for pushing it to the top of my priority list and keeping it there until done.

A Necessary Prelude

x   **In 1986 I introduced the world to the now well-known concept of nanotechnology:** At the outset, and as widely understood today (though other meanings have taken hold), "nanotechnology" refers to large-scale, atomically precise manufacturing based on the principle of mechanically directed placement of molecules at the atomic scale. Both the word and the principle had precursors.

Regarding the physical principle behind the concept, Nobel Prize–winning physicist Richard Feynman had earlier suggested that machines operating at this scale should be feasible and could indeed be used to direct atomically precise fabrication. He suggested this in a talk delivered in 1959, but the idea lay fallow for almost twenty years.

Regarding the word itself, "nanotechnology" was an obvious yet ambiguous coinage patterned on "microtechnology"; I later learned that Norio Taniguchi had once used essentially the same term ("Nano-Technology") in a conference paper to refer to ultra-high-precision machining and the like, a quite different range of technologies.

(Richard Feynman, "There's Plenty of Room at the Bottom," *Engineering and Science* 23, no. 5 [1960]; Norio Taniguchi, "On the Basic Concept of 'Nano-Technology'," *Proceedings of the International Conference of Production Engineering*, Vol. 2. [Tokyo: Japan Society of Precision Engineering, 1974].)

x   **Nanoscale parts and atomic precision together enable atomically precise manufacturing:** See Chapter 10 and Appendix I. An Institute of Physics journal provides a tutorial overview of the topic (K. Eric Drexler, "Productive Nanosystems: The Physics of Molecular Fabrication," *Physics Education* 40 [2005]: 339).

xi   **Much of the most important research is seldom called "nanotechnology":** Chapter 12 surveys the status and rapid progress in the technologies of atomic precision, while the following chapter tells the story of how atomic precision and (federally funded) nanotechnology diverged.

Chapter 1: Atoms, Bits, and Radical Abundance

5   **to retrieve images of pages stored on microfilm:** Bush's proposed "memex" system would have been more than that, however; he proposed what amounted to a pre-digital version of a hypertext system.

7   **In mechanically guided chemical processes:** Appendix I discusses the physical principles and requirements.

Chapter 2: An Early Journey of Ideas

9   **a scientific paper I published in 1981:** Cited in the main text and available at www .pnas.org/content/78/9/5275.full.pdf+html. Although published in the *Proceedings of the National Academy of Sciences*, this paper stands on the engineering side of the science/ engineering distinction drawn in Chapter 8.

10   **book-length analysis based on my MIT dissertation:** "Molecular Machinery And Manu-facturing With Applications To Computation" (1991), completed in an interdepartmental doctoral program in the field of Molecular Nanotechnology. I started the project (as a book) while teaching a seminar at Stanford University in 1988, and after a further year of expansion and revision the dissertation turned into a book project again and became *Nanosystems: Molecular Machinery, Manufacturing, and Computation* (Hoboken, NJ: Wiley/Interscience, 1992). The dissertation is online at http://dspace.mit.edu/handle/1721.1/27999 and available at e-drexler.com/d/09/00/Drexler_MIT_dissertation.pdf. A detailed table of contents and sample chapters for *Nanosystems* is available at e-drexler.com/d/06/00/Nanosystems/toc.html.

11   **This vision for the human future:** The leading visionaries included Freeman Dyson, Gerard O'Neil, Dandridge Cole, J. D. Bernal, and the father of theoretical astronautics, Konstantin Tsiolkovsky.

15fn1   **missions to asteroids have become part of NASA's plans:** See, for example, http:// www.nasa.gov/about/obamaspeechfeature.html.

15fn2   **because in the end Malthus was right:** In his *Essay on the Principle of Population*, Reverend Thomas Malthus argued that population tended to grow exponentially and that this expo-nential growth would overrun the limits of food production even if production steadily in-creased. His general argument regarding exponential growth holds true for any imaginable production technology within the bounds of the material universe.

18   **His bold visions started early:** In our forward-looking conversations Arthur had occasion to mention only a few of his past accomplishments. The sample I list draws on a document I discovered while writing this chapter ("Oral History Transcript—Dr. Arthur Kantrowitz," Center for History of Physics of the American Institute of Physics, http://www.aip.org /history/ohilist/31816.html).

20fn   **reading journals like *Science* and *Nature*:** The best way of grasping the structure and content of modern science is to read journals like these, year after year, until almost everything one encounters fits into a framework of knowledge learned much like a natural language, through context and use. Formal, focused study of fundamentals (starting with mathematics and physics) provides a necessary complement.

Chapter 3: From Molecules to Nanosystems

24   **Feynman proposed the idea of using machine-guided motion to assemble molecular structures with atomic precision:** "The principles of physics, as far as I can see, do not speak against the possibility of maneuvering things atom by atom . . . it would be, in principle, possible (I think) for a physicist to synthesize any chemical substance that the chemist writes down. Give the orders and the physicist synthesizes it. How? Put the atoms down where the chemist says, and so you make the substance" (Richard Feynman, "There's Plenty of Room at the Bottom," *Engineering and Science* 23, no. 5 [1960], http://www.its.caltech.edu /~feynman/plenty.html).

         "Put the atoms down where the chemist says" isn't quite right, however—a chemist would think in terms of bonding interactions between molecular structures; atomic *precision* does not entail atom-by-atom *construction*.

27fn   **widely cited in the scientific literature as a foundation:** Carl Pabo dubbed the 1981 paper's key concept "inverse folding" (C. Pabo, "Molecular Technology: Designing Proteins and Peptides," *Nature* 301 [1983]: 200), and Jay Ponder and Frederick Richards took this idea further, providing a pivotal insight for practical implementation (J. W. Ponder and F. M. Richards, "Tertiary Templates for Proteins: Use of Packing Criteria in the Enumeration of Allowed Sequences for Different Structural Classes," *Journal of Molecular Biology* 193 [1987]: 775–791). An overview of the field at a later stage of development also notes some of the history (O. Alvizo, B. D. Allen, and S. L. Mayo, "Computational Protein Design Promises to Revolutionize Protein Engineering," *BioTechniques* 42, no. 1 [2007]: 31–37).

29   **Scientific inquiry long ago uncovered the fundamental principles of molecular physics:** E. Schrödinger, "Quantisierung als Eigenwertproblem," *Annalen der Physik* 385, no. 13 (1926): 437–490.

30   **spurred studies that deepened our understanding of the prospects:** Ralph Merkle was the most notable early contributor; I collaborated with him at Xerox Palo Alto Research Center while I wrote *Nanosystems* and he later helped to guide the molecular manufacturing research program funded by the NASA Ames Research Center until the federal-level upheaval documented in Chapter 13 put an end to such things.

31   **it seemed that enthusiasm was primarily a positive force:** In 1986, I cofounded the Foresight Institute (later renamed the Foresight Nanotech Institute); the institute became a center for nanotechnology enthusiasts and also sponsored a leading scientific conference series in the field for the following decade or so. I served as a largely hands-off chairman until shortly after 2000, and a few years later I severed my relationship.

31   **researchers at Caltech and elsewhere applied computational methods:** This work was supported by the NASA funding mentioned above.

32   **scanning probe instruments to image and place individual atoms:** For example, Y. Sugimoto et al., "Complex Patterning by Vertical Interchange Atom Manipulation Using Atomic Force Microscopy," *Science* 322, no. 5900 (17 October 2008): 413–417.

32   **to maneuver and bond individual molecules:** This level of control illustrates the principle of mechanically directed atomically precise fabrication (see, for example, S. W. Hla and K. H. Rieder, "STM Control of Chemical Reactions: 'Single-Molecule Synthesis,'" *Annual Review of Physical Chemistry* 54 (2003): 307–330.

32   **Protein engineering . . . supported by computer-aided design software:** A landmark paper was "Tertiary templates for proteins. Use of packing criteria in the enumeration of allowed sequences for different structural classes," J. W. Ponder and F. M. Richards, *Journal of Molecular Biology* 193, no. 4 (1987): 775–791; dx.doi.org/10.1016/0022-2836(87)90358-5.

33   **Quantum methods . . . powerful, physics-based tools:** Reviewed in Trygve Helgakera, Wim Klopper, and David P. Tew, "Quantitative Quantum Chemistry Preview," *Molecular Physics*, 106, issues 16–18 (2008): 2107–2143.

33   **Molecular mechanics methods in chemistry:** These methods can be used in conjunction with quantum chemistry methods in several ways, including their integration in directly applied computations. A particularly challenging range of applications (involving the calculation of small free-energy differences that depend in part on entropy) is described here: "Free Energies of Chemical Reactions in Solution and in Enzymes with Ab Initio Quantum Mechanics/Molecular Mechanics Methods," *Annual Review of Physical Chemistry* 59 (May 2008): 573–601. First published online as a Review in Advance on December 11, 2007 (dx.doi.org/10.1146/annurev.physchem.59.032607.093618) by Hao Hu and Weitao Yang.

33   **the APM-based production revolution:** The main discussion of the technological basis appears in chapters 10 and 11, supported by Appendix I.

Chapter 4: Three Revolutions, and a Fourth

41   **Markers in soil and genes show the spread of a wave of farmers:** P. Skoglund, et al. "Origins and Genetic Legacy of Neolithic Farmers and Hunter-Gatherers in Europe," *Science* 336, no. 6080 (2012): 466–469.

47   **led them to invent the world's first transistor:** Today's digital logic systems are based on field-effect transistors (J. Bardeen and W. H. Brattain, "Three Electrode Circuit Element

Utilizing Semiconductive Materials," U.S. Patent 2,524,035 [1950]) which had been anticipated—based on physics, not experimentation—twenty years before (J. E. Lilienfeld, "Method
and Apparatus for Controlling Electric Currents," U.S. Patent 1,745,175 [1930]).

48   **"Moore's Law" and stated that every two years:** The meaning of "Moore's Law" has shifted
and broadened over time, and a similar pattern of exponential progress is seen in areas as
different as optical fiber data rates and bit-densities in magnetic storage media. In the molecular domain gene-sequencing technologies have advanced on an even steeper exponential
curve.

Chapter 5: The Look and Feel of the Nanoscale World

55   **These are the solid, rigid molecular objects:** In the context of APM-level nanomechanical
components, the typical molecules of interest are stable covalent structures that consist of
fused rings; among the hydrocarbons, small-scale examples include the adamantanes and
the somewhat more flexible polycyclic aromatic molecules. Lacking conformational degrees
of freedom, these structures maintain definite shapes, and bending them introduces substantial strain energy, hence elastic mechanical restoring forces.

Many other molecules, by contrast, are highly flexible. Polymers, for example, are often
modeled as jointed, swiveling chains. The molecules of practical interest for nanomechanical
engineering today are polymeric chains that fold to form compact structures; these have
definite shapes, while molecules of greater rigidity pose greater challenges to solution-phase
synthesis.

56   **rigid molecular structures:** In other words, the quantum fluctuations associated with the
nuclear position coordinates are small compared to thermal fluctuations at ambient temperatures, which in turn are small compared to bond lengths and atomic diameters. The
dominance of thermal over quantum fluctuations is one reason for the success of classical
mechanics in describing molecular dynamics.

56   **gain a visual, even kinesthetic, understanding:** A haptic interface to a computational model
can provide a literal kinesthetic understanding.

62   **the nanoscale structure of a gasoline engine:** To get a sense of what precision machining
looks like from a nanoscale perspective, consider a ball bearing at our standard magnification:
A one-millimeter ball would be a sphere ten kilometers in diameter, while a ball with a finely
polished, Grade-3 surface would display meter-scale gouges.

66   **experimentalists have demonstrated each of these principles:** Alex Zettl's lab has led in
this area. For a review of computational and experimental work with nanomechanical bearings of this kind, see A. Kis and A. Zettle, "Nanomechanics of Carbon Nanotubes," *Philo-
sophical Transactions of the Royal Society A* 366, no. 1870 (2008): 1591–1611 (http://
rsta.royalsocietypublishing.org/content/366/1870/1591.full).

66   **This is a drag force, like the resistance:** Drag forces result from both scattering and radiation
of phonons from the sliding interface. Because these forces typically scale in proportion to
velocity, the energy dissipated by a bearing surface in sliding a given distance is proportional
to the motion frequency and inversely proportional to the motion distance. For nanoscale
motions, energy losses can be low even at frequencies in the 100 MHz range.

71fn   **the method of choice for molecular assembly today:** Self-assembly is practical and effective,
and at present, alternative methods are seldom available.

Chapter 6: The Ways We Make Things

79   **discreteness is inherent . . . in the very nature of typical covalent bonds:** This is true of a
large but not comprehensive range of structures; others (often best avoided) have less distinct
patterns of bonding that may be subject to rapid thermal fluctuations. More generally, the
key criterion isn't bonding per se, but the magnitudes of energy barriers between different
configurations.

80   **Well-chosen APM operations, guided by well-designed mechanisms:** Appendix I outlines
some of the requirements. These include favorable reaction kinetics and thermodynamics,

adequate mechanical stiffness, and substantial geometrical separations between transition states leading to alternative reaction outcomes; each of these requirements stems from the need to constrain thermal fluctuations.

82   **The methods of organic chemistry aren't widely understood:** In his 1959 talk, Richard Feynman joked that "The chemist does a mysterious thing when he wants to make a molecule. He sees that it has got that ring, so he mixes this and that, and he shakes it, and he fiddles around. And, at the end of a difficult process, he usually does succeed in synthesizing what he wants." (Richard Feynman, "There's Plenty of Room at the Bottom," *Engineering and Science* 23, no. 5 [1960].)

Chapter 7: Science and the Timeless Landscape of Technology

90   **a timeless aspect of physical law, a latent structure:** Scientific understanding of physical law has changed over time, but physical law itself has not, and within the accuracy of both laboratory measurement and cosmological observation, "physical constants" do indeed appear to be constant.

96   **Over the last four hundred or so years:** Sean Carroll, "Cosmic Variance" blog at Discover Magazine, 16 June 2010 (http://blogs.discovermagazine.com/cosmicvariance/2010/06/16/reluctance-to-let-go/).

Chapter 8: The Clashing Concerns of Engineering and Science

107   **Engineering:** The word "engineering" derives from "ingenium," the Latin root of "ingenuity." Locomotives and lawnmower engines aren't what engineering is about.

108   **In the story as the Jains tell it:** Fabien Schang, "A Plea for Epistemic Truth: Jaina Logic from a Many-Valued Perspective," in *Logic in Religious Discourse*, edited by Andrews Schumann (Piscataway, NJ: Transaction Books, 2010), 54–83.

113   **At the deepest, epistemic level, scientific inquiry:** There are extensive academic literatures on the philosophy of science and (less so) on the philosophy of engineering. The view I describe here is neither entirely fresh, nor entirely old. I will leave it to others more familiar with these literatures to sort out what may be novel.

114   **But the close, intertwining links:** The structural contrasts between science and engineering can be laid out as follows:

Domain:	Scientific inquiry	Engineering design
Basic purpose:	provides knowledge	provides products
Information flow:	from world to model	from model to world
Ideal models:	exact descriptions	reliable bounds
Desired results:	surprising discovery	predictable behavior
Strongest knowledge:	excludes all alternatives	includes many alternatives
Effective organization:	independent groups	coordinated teams

118fn   **the BBC reported:** "North Korea 'installs long-range rocket at launch site,'" 3 December 2012 (http://www.bbc.co.uk/news/world-asia-20577340).

Chapter 9: Exploring the Potential of Technology

133   **The vision was spaceflight and the man was Konstantin Tsiolkovsky:** A. deChambeau, "Struggles of the 'Father,'" *Ad Astra* (September/October 2002): 41–44 (http://dspace.sunyconnect.suny.edu/bitstream/handle/1951/36567/deChambeau-AdAstra.pdf) and the Tsiolkovsky State Museum of the History of Cosmonautics (http://www.gmik.ru/index_en.html).

       Konstantin Eduardovich Tsiolkovsky, *Beyond the Planet Earth* (1920), page 13. From the foreword by B. N. Vorobyev. Translated by Kenneth Syers. Pergamon Press, 1960. These words are inscribed on the obelisk that marks his grave: "Man will not always stay on Earth;

the pursuit of light and space will lead him to penetrate the bounds of the atmosphere, timidly at first but in the end to conquer the whole of solar space."

139   **Engineering with an Exploratory Twist:** The structural contrasts between conventional and exploratory engineering can be laid out as follows:

Kind of engineering:	Production-oriented	Exploratory
Basic purpose:	provides products	provides knowledge
Basic constraint:	accessible fabrication	valid modeling
Level of design:	detailed specification	parametric model
Primary costs:	production, operation	design, analysis
Design margins:	enable robust products	enable robust analyses
Larger margins:	increase costs	reduce costs

Chapter 10: The Machinery of Radical Abundance

150   **successive layers will have thicknesses of 1, ½, ¼, ⅛, and so on:** This neat, self-similar architecture was suggested by Ralph Merkle and improves on the functionally similar version described in *Nanosystems.*

153   **inherently messy contact with the stuff of nature:** In converting raw materials into purified feedstocks, the path to atomic precision leads through reducing molecular complexity, a natural task for conventional chemical processes like those used in industrial processes to dissolve minerals and convert their materials into simple molecular and ionic species. These can then be separated through cascade sorting processes well suited to atomically precise mechanisms, as discussed in *Nanosystems.*

153   **with a site that binds a feedstock molecule:** A mechanism downstream can probe each site to ensure that it's full and push empties on a path that leads them around for another go. With an iterative process of this kind, the free energy requirement for reliable binding can approach the minimum required for reducing entropy, which can be thought of as the work required for compression in a configuration space with both positional and angular coordinates.

     Computational chemists will note that free energy calculations involving oriented, non-solvated molecules in a rigid binding site are far less challenging than calculations that must probe the configuration space of mobile, solvated, conformationally flexible molecules. The former frequently have only vibrational degrees of freedom, sometimes enabling the use of analytic approximations to the entropy and free energy.

154   **the full range of materials used in technology today:** Any exceptions will likely be of only academic interest, and such materials could still be made by more conventional means. Adding APM to the engineering toolkit won't preclude the use of other means that remain superior.

154   **The downstream mechanisms guide molecules through a series of encounters:** In current industrial practice continuous-motion machines are widely used to assemble streams of products from streams of parts; both streams flow smoothly as parts mesh together, without the need for computing or for motors to push arms to and fro. The springs and bits of plastic in a hand-pumped spray bottle, for example, aren't assembled by human or robotic hands; they simply flow through machines with rotary, sliding, and cam-driven components that orient and bring parts together at rates of up to ten per second. (Web videos of continuous motion assembly machines show the principles.)

     Analogous mechanisms are the natural choice for bringing monomers and small parts together on the way to building microblocks. Scaled by the ten million to one ratio (*per* Chapter 5), a rate of ten operations per second performed by macroscale mechanisms corresponds to a rate of one million operations per second performed by nanoscale mechanisms. A large slowdown factor from this base (to reduce phonon drag in bearings, for example) is compatible with a still-enormous product throughput.

154   **chemical steps that prepare reactive bits of molecular structure:** To work reliably with small reactive groups and fine-grained structures, placement mechanisms must be stiff

enough to adequately constrain thermal fluctuations. (Appendix I and Appendix II place this requirement in perspective.)

156 **each step typically must expend substantial chemical energy:** As with binding, an iterated process with conditional repetition can in some instances avoid this constraint.

157 **density functional methods . . . applied in conservative ways:** Methods in quantum chemistry have limited accuracy and ranges of applicability that must be kept in mind when considering which methods to use and how far to trust their results. Density functional methods, for example, typically underestimate the energies of reaction transition states, and this may or may not be acceptable, depending on the intended application. (Where high barriers are desirable, e.g., to maintain structural stability, underestimates are conservative.)

Chapter 11: The Products of Radical Abundance

162 **structures made of stronger, lighter materials:** Although the strongest materials are also brittle (as a consequence of their strong, directional bonding), they can be incorporated into robust materials in fibrous form, like glass in glass-fiber composites. Fibrous AP materials can be optimized to provide extreme fracture toughness based on the fiber-pullout mechanism in composite materials, together with what amounts to work-hardening that is tailored to ensure a growing deformation zone rather than crack propagation.

162 **A lightweight material as stiff as aircraft aluminum:** Young's modulus for stiff carbon materials exceeds that of conventional metals by a large ratio (relative to aluminum, a factor of fifteen), but this is less than the strength ratio. Elastic modulus is often more important than strength in structures subject to bending and compressive loads, but advanced AP materials and structures offer two compensating advantages: High elastic strain limits (e.g., 10 percent) allow greater deflection without structural failure, while the ease of producing structural components with full-density skins and low-density cores facilitates the use of structures that are simultaneously lighter and thicker (hence more resistant to buckling). In vehicle structures designed to exploit these characteristics, the large strength-ratio advantage of carbon materials should typically enable a similar mass-ratio advantage.

164 **improved photovoltaic cells for solar energy:** The requirements for photovoltaic cells based on Earth-abundant materials are well understood; the challenge today is one of low-cost, low-defect-density fabrication (C. Wadia, A. P. Alivisatos, and D. M. Kammen, "Materials Availability Expands the Opportunity for Large-Scale Photovoltaics Deployment," *Environmental Science and Technology* 43, no. 6 [2009]: 2072–2077. http://pubs.acs.org/doi/abs /10.1021/es8019534).

164 **electromagnets don't scale according to mechanical rules:** Holding current density constant, the magnetic field energy density adjacent to an electrical conductor—hence the force per unit area—varies in direct proportion to diameter, while in mechanical scaling the force per unit area is constant. Thus, with a scale reduction of ten million, the ratio of magnetic to mechanical forces decreases by ten million.

At constant voltage, by contrast, electric field forces per unit area increase as the square of this ratio (halving the distance doubles the field, quadrupling its energy density). Holding field strength constant (a requirement when fields approach breakdown strength), electrical forces scale in proportion to mechanical forces. Low voltage devices operating with high electric fields turn out to be quite attractive as nanoscale motors (as discussed in more detail in *Nanosystems*). Efficiencies can be high, and scaling laws enable extraordinarily high power densities.

165 **This hugely exceeds the power density of today's:** In nanoscale systems, thermal path lengths become short and small temperature differences result in a high thermal energy flux (the ratio goes as the scaling factor, e.g., ten million times greater). In macroscale aggregates of nanoscale devices, however, thermal path lengths remain ordinary and opportunities for improved cooling are modest.

165 **mechanical scaling laws enable systems with:** High throughput in APM-based materials processing and high throughput in AP systems–based energy conversion both originate in the properties of aggregates of small, high-frequency devices. High efficiency depends (in

part) on the use of low-entropy, non-thermal processes and mechanisms with low internal friction.

165   **downhill steps . . . coupled tightly to mechanical motions:** To make such processes nearly thermodynamically reversible requires a combination of high stiffness in the mechanical systems (in particular, along their motion coordinates) and low negative curvature of the potential energy function (in mechanical terms, low negative stiffness) along transition-state reaction coordinates. The required conditions are stringent, but can apparently be met by making use of transition metal catalysis and stiff covalent components supported by stiff, compressed bearing interfaces.

165   **enough to enable nanomechanical computers to compete:** *Nanosystems* includes a detailed analysis. Note that the speed of mechanical signal propagation in stiff carbon materials is on the order of ten kilometers per second, while the diameter of a reasonably capable computer processor core can be on the order of a micron, implying signal propagation delays on the order of 0.1 nanosecond.

167   **device efficiencies closer to the thermodynamic limits:** Again, *Nanosystems* includes a detailed analysis.

167   **in the range of a billion gigabytes per cubic centimeter:** This is a conservative value for the storage density of a class of static RAM technologies. Higher-latency devices further down in the memory hierarchy can store data at over one thousand times this density, based on polymers less bulky than DNA.

169   **As for scarcer elements like zinc, tin, and lead:** With respect to materials substitution, the primary areas in which scarce materials may hold a continuing advantage involve electronic, magnetic, and chemical properties. For example, the leading high-temperature superconductors today incorporate not only copper, but elements like yttrium, barium, bismuth, and strontium. Today, however, commercial applications are rare.

For another example, neodymium and other rare-earth metals are currently scarce and are required for magnets used in high-performance electric motors for electric automobiles and wind turbines. For fundamental reasons of electronic structure (the number of unpaired spins per atom), rare-earth elements will likely remain superior for use in permanent-magnet materials, but for reasons suggested above, magnet-based motors themselves will likely be superseded.

In another example, platinum and other rare transition metals have unique properties as components of catalysts. Here, however, mechanically directed chemical processes have a remarkable implication. Catalysts incorporated into high-frequency, AP mechanisms can have both long lifetimes and turnover frequencies that exceed one million per second (and can be recycled, of course). In this use, the quantities of rare transition metals required become very small indeed.

Adding up costs in a summary overview:

Source of cost component	Typical ratio	Reason for cost reduction
Raw materials	< 1/10	Low product masses, low-cost raw materials
Energy inputs	~ 1	Low product masses offset increments in per-mass energy cost
Land required	< 1/100	Compact high-throughput process equipment
Labor inputs	< 1/100	No direct process intervention
Waste disposal	< 1/100	Low mass processed, precise process control, zero carbon footprint
Accidents	< 1/100	No hazardous materials or direct human exposure
Physical capital	< 1/1	High-throughput, low cost production equipment

With a wide range of performance improvements relative to current products and a wide range of current dollar-per-kilogram production costs, cost-reduction ratios of 1/10 to 1/1000 can be expected.

Chapter 12: Today's Technologies of Atomic Precision

179   **Chemists make compounds/molecules, the objects of their own contemplation:** R. Hoffmann, "What Might Philosophy of Science Look Like If Chemists Built It?" *Synthese* 155, no. 3 (2007): 321–336 (http://www.springerlink.com/content/px87858157672k37/).

180   **In a landmark 2001 paper:** H. C. Kolb, M. G. Finn, and K. B. Sharpless, "Click Chemistry: Diverse Chemical Function from a Few Good Reactions," *Angewandte Chemie International Edition* 40, no. 11 (2001): 2004–2021.

181   **The Chemical Abstracts Service:** September 2012.

181   **peptoids, made by an extraordinarily convenient method:** R. N. Zuckermann et al., "Efficient Method for the Preparation of Peptoids [oligo(N-substituted glycines)] by Submonomer Solid-Phase Synthesis," *Journal of the American Chemical Society* 114, no. 26 (1992): 10646–10647; and an invited overview of a recent peptoid conference: K. E. Drexler, "Peptoids at the 7th Summit: Toward Macromolecular Systems Engineering," *Peptide Science* 96, no. 5 (2011): 537–544.

182   **Researchers sometimes design proteins from scratch:** A review: Robert J. Pantazes, Matthew J. Grisewood, and Costas D. Maranas, "Recent Advances in Computational Protein Design," *Current Opinion in Structural Biology* 21, no. 4 (2011): 467–472 (http://dx.doi.org/10.1016/j.sbi.2011.04.005).

183   **read a code with four—not three—bases per codon:** K. Wang, W. H. Schmied, and J. W. Chin, "Reprogramming the Genetic Code: From Triplet to Quadruplet Codes," *Angewandte Chemie International Edition* 51, no. 10 (2012): 2288–2297.

184   **Starting in the 1980s, Nadrian Seeman:** One of his landmark papers is J. Chen and N. C. Seeman, "Synthesis from DNA of a Molecule with the Connectivity of a Cube," *Nature* 350 (1991): 631–633.

184   **a technique called "DNA origami":** More specifically, *scaffolded* DNA origami: P. W. K. Rothemund, "Folding DNA to Create Nanoscale Shapes and Patterns," *Nature* 440 (2006): 297–302.

185   **nudged thirty-five xenon atoms:** Donald M. Eigler and Erhard K. Schweizer, "Positioning single atoms with a scanning tunnelling microscope," *Nature* 344, no. 6266 (1990): 524–526.

187   **Putting the Pieces Together: Composite Molecular Systems:** These include early-stage products along paths outlined in Appendix II.

188   **DNA does have shortcomings:** Aptamers are molecules that fold and bind to non-DNA structures. DNA machines include, for example, mechanisms with sets of swinging arms (Baoquan Ding and Nadrian C. Seeman, "Operation of a DNA Robot Arm Inserted into a 2D DNA Crystalline Substrate," *Science* 314, no. 5805 [2006]: 1583–1585.) together with "walkers" and other devices (Jonathan Bath and Andrew J. Turberfield, "DNA Nanomachines," *Nature Nanotechnology* 2, no. 5 [2007]: 275–284).

189   **it will soon be extended to peptoids:** Kent Kirshenbaum, personal communication.

190   **leaves design as the primary problem:** Regarding engineering and the limited predictive power of current computational models, two points should be kept in mind: First, the (approximate) additivity of interaction energies can make larger structures less sensitive to errors in calculated energies and, second, that trial and error has always been part of the engineering design.

Chapter 13: A Funny Thing Happened on the Way to the Future . . .

194   **Dr. Abdul Kalam . . . delivered a series of addresses:** For example, Dr. Kalam's address to scientists and technologists in April 2004 in Delhi (adapted for the Hindustan Times, 25/11/2004: GUEST COLUMN, President A. P. J. Abdul Kalam, http://www.hindustantimes.com/News-Feed/NM3/President-Kalam-s-Take/Article1-21101.aspx) and his recent "Address at the Dr Brahma Prakash Birth Centenary Celebrations," Hyderabad, November 30, 2012.

194   **More than a decade ago . . . President Clinton announced a plan:** In a science policy address at Caltech, January 21, 2000: "My budget supports a major new National Nanotechnology Initiative, worth $500 million. Caltech is no stranger to the idea of nanotechnology—the

ability to manipulate matter at the atomic and molecular level. More than forty years ago, Caltech's own Richard Feynman asked, 'What would happen if we could arrange the atoms one by one the way we want them?'"

194   **China's program has a larger (and rapidly growing) budget:** "Public Funding of Nanotechnologies 2012," Report into Global Public Funding of Nanotechnologies, ObservatoryNANO project (7th Framework programme), www.observatorynano.eu/project/filesystem /files/PublicFundingofNanotechnologies_March2012.pdf. "China Triples Spending on Nanotechnology over Past Five Years," Xinhua Global Edition (English), January 2010 (http://news.xinhuanet.com/english2010/china/2011-01/11/c_13686054.htm).

194   **has reached a billion dollar scale in the United States, European Union, and China:** "Global Funding of Nanotechnologies and Its Impact, July 2011" Cientifica (http://cientifica.com/wp-content/uploads/downloads/2011/07/Global-Nanotechnology-Funding-Report-2011.pdf).

197   **"nanotechnology" was simply a name I had chosen:** As mentioned in a note to the Prelude, I later learned of a use of "Nano-Technology" in a 1974 Japanese conference paper on a different range of technologies.

197   **the popular press reached millions of readers:** Starting with *OMNI* in late 1986, as noted in Chapter 3.

198fn   **a title with the words "Iron Oxide Nanoparticles":** *Science* 4 (May 2012): Ajay Kumar Gupta, and Mona Gupta. "Synthesis and Surface Engineering of Iron Oxide Nanoparticles for Biomedical Applications," *Biomaterials* 26, no. 18 (2005): 3995–4021.

199   **"the ones that will clean out our arteries?":** The popular image of microscopic submarines swimming among red blood cells and chomping fat deposits began with artwork produced by the staff of *Scientific American* to illustrate an article in the "Computer Recreations" section at the end of the magazine (A. K. Dewdney, "Nanotechnology: Wherein Molecular Computers Control Tiny Circulatory Submarines," *Scientific American* 258 [January 1988]: 100–103 [101]).

203   **embraced the rhetoric of "building atom by atom":** The idea of literally building "atom by atom" is itself a technically inaccurate popularization of the idea of atomically precise fabrication—chemical processes routinely yield atomically precise results, yet never juggle individual atoms. Ironically, the popular idea that APM would require impossible atom juggling became a popular criticism among scientists who seemingly neither read the literature nor considered how APM might actually be accomplished.

204   **a glossy "brochure for the public," entitled:** *Nanotechnology: Shaping the World Atom by Atom,* Interagency Working Group on Nanoscience, Engineering and Technology (December 1999) (http://www.wtec.org/loyola/nano/IWGN.Public.Brochure/).

204   **and the NSTC issued a more formal document:** *National Nanotechnology Initiative: The Initiative and Its Implementation Plan,* National Science and Technology Council Committee on Technology Subcommittee on Nanoscale Science, Engineering and Technology (July 2000) (http://nano.gov/sites/default/files/pub_resource/nni_implementation_plan_2000.pdf).

205   **In the section titled "Definitions":** Sec. 8., Definitions, 21st Century Nanotechnology Research and Development Act, 107th Congress, 2001–2002 (the same definition appears in a series of subsequent bills).

205fn   **a bill was introduced that would strike:** Sec. 13., Amendments to Definitions, National Nanotechnology Initiative Amendments Act of 2008, 110th Congress, 2007–2009.

206   **the 2004 *National Nanotechnology Initiative Strategic Plan*:** National Science and Technology Council Committee on Technology Subcommittee on Nanoscale Science, Engineering and Technology (December 2004) (www.nsf.gov/crssprgm/nano/reports/sp_report_nset _final.pdf).

207   **published an article on future technologies in *Wired* magazine:** Bill Joy, "Why the Future Doesn't Need Us," Issue 8.04 (April 2000): 238–262.

208   **The principal fear is that it may be possible to create:** R. E. Smalley, "Nanotechnology, Education, and the Fear of Nanobots," in *Societal Implications of Nanoscience and Nanotechnology,* a report from a National Science Foundation workshop on September 28–29, 2000 (www.wtec.org/nanoreports/nanosi.pdf). Here and in *Scientific American,* Smalley repeated, elaborated, and reinforced misconceptions that came out of popular fiction.

208 **equated APM ... with swarms of dangerous nanobots:** R. E. Smalley, "Of Chemistry, Love and Nanobots," *Scientific American* 285, no. 3 (September 2001): 76–77 (cohesion .rice.edu/naturalsciences/smalley/emplibrary/sa285-76.pdf). I give particular attention to Smalley's statements here and below because, as a Nobel Prize winner and the most prominent scientific spokesman for the NNI program, his views carried great weight in the controversies of the time and he was widely regarded as the leading critic of what were wrongly said to be my views.

Note that, in an article in the same issue, I had stated that a "well-known chemist objects that an assembler [a device for stereotactic synthesis] would need ten robotic 'fingers' to carry out its operations and that there wouldn't be room for them all," a requirement that "has never been established or even seriously argued" (in K. E. Drexler, "Machine-Phase Nanotechnology," *Scientific American* 285, no. 3 [September 2001]: 74–75). Nonetheless, just a few pages later, Smalley claimed once again that atomically precise manufacturing would require tiny-but-impossible fingers, and his remarks were widely cited as both explaining and critiquing how APM systems must work.

208 **in his congressional testimony and other statements:** Smalley's alternating positions regarding atomic precision: "We are about to be able to build things that work on the smallest possible length scales, atom by atom" (in testimony to the Committee on Science, US House of Representatives, June 22, 1999); "To put every atom in its place—the vision articulated by some nanotechnologists—would require magic fingers" (in *Scientific American*, September 2001, cited above); "The ultimate nanotechnology builds at the ultimate level of finesse one atom at a time, and does it with molecular perfection" (in a presentation to the President's Council of Advisors on Science and Technology, March 3, 2003). For several years, Smalley's second, negative opinion was widely cited as authoritative, but only in connection with (what were said to be) proposed APM technologies.

208 **acknowledged as inspiring his enthusiasm for nanotechnology:** "Reading [*Engines of Creation*] was the trigger event that started my own journey in nanotechnology," in "Nanotechnology: Drexler and Smalley make the case for and against molecular assemblers," Part Two: "Smalley Responds," *Chemical and Engineering News*, 81 (2003): 39–40.

209 **Smalley subsequently spoke out against Darwin:** "Scholarship Convocation Speaker Challenges Scholars to Serve the Greater Good," Tuskegee University, October 3, 2004, News Release: "Smalley mentioned the ideas of evolution versus creationism, Darwin versus the Bible's Genesis. The burden of proof, he said, is on those who don't believe that 'Genesis was right, and there was a creation, and that [the] Creator is still involved.'"

210 **a bill directing the National Academy of Sciences:** H.R. 766 (108th) Nanotechnology Research and Development Act of 2003, 108th Congress, 2003–2004. Text as of May 8, 2003 (Referred to Senate Committee).

211 **lobbied ... to strike the request from the bill:** "In fact, another NanoBusiness Alliance official had already admitted to a reporter that the Alliance had approached the staff of Senator John McCain, Republican of Arizona, to have the study removed from the legislation [in the House-Senate conference committee]." (The editors of *The New Atlantis*, "The Nanotech Schism," *The New Atlantis*, Number 4 [Winter 2004]: 101–103 [cited in http://www.the newatlantis.com/publications/the-nanotech-schism].)

211 **organized a roadmap project:** *Productive Nanosystems: A Technology Roadmap*, K. E. Drexler, J. Randall, and A. Kawczak, editors, Battelle Memorial Institute, 2007 (http://productivenanosystems.com/pgs/Nanotech_Roadmap_2007_TOC.html).

Chapter 14: How to Accelerate Progress

215 **Roadmapping Moore's Law:** The International Technology Roadmap for Semiconductors (http://www.itrs.net/Links/2011ITRS/Home2011.htm).

215 **Roadmapping Quantum Information Systems:** The Quantum Information Science and Technology Roadmap (http://qist.lanl.gov/qcomp_map.shtml).

216 **Roadmapping Paths to APM:** *Productive Nanosystems: A Technology Roadmap*, K. E. Drexler, J. Randall, and A. Kawczak, editors, Battelle Memorial Institute, 2007 (http://productivenanosystems.com/pgs/Nanotech_Roadmap_2007_TOC.html).

Chapter 15: Transforming the Material Basis of Civilization

226   **the limits are already in sight:** In flash memory we see a clear sign of a break with Moore's Law progress, which has been based on shrinking devices. Manufacturers have begun to re-sort to stacking layers to add devices, but this adds an increment in processing cost with each layer. Unlike shrinking the devices themselves, stacking layers isn't a recipe for expo-nential improvements.

226   **from milliwatts to nanowatts:** Through technologies that operate substantially closer to (yet still short of) the ultimate thermodynamic limits for digital computation (C. H. Bennett, "Notes on the History of Reversible Computation," *IBM Journal of Research and Development* 32 [1988]: 16–23).

229   **ten-fold reductions in vehicle mass:** As a consequence of improvements in the strength of practical structural materials (discussed in Chapter 11).

229   **a doubling of typical engine efficiencies:** See the discussion of chemical energy conversion in Chapter 11 and its notes; automobile and truck engines today are substantially less than 50 percent efficient.

230   **coal-fired power plants (2,300 today):** International Energy Agency Clean Coal Centre, http://www.iea-coal.org.uk/site/2010/database-section/coal-power (retrieved December 2012).

230   **By enabling abundant elements . . . to substitute for scarcer materials:** Chapter 11 (and notes) describe some of the main aspects of this materials-substitution opportunity.

231   **Atomically precise fabrication can produce membranes:** For example, short carbon nano-tubes and nanoporous graphene provide models for flow pathways through low-resistance membranes (e.g., D. Cohen-Tanugi and J. C. Grossman, "Water Desalination across Nanoporous Graphene," *Nano Letters* 12, no. 7 [2012]: 3602–3608).

231   **food prices have recently trended sharply upward:** The FAO food price index shows prices trending upward since 2000 and are currently roughly double those of the 1990s. Opinions regarding future trends are mixed, as usual.

238   **converting scar-tissue cells into cardiac muscle:** L. Qian et al., "*In vivo* Reprogramming of Murine Cardiac Fibroblasts into Induced Cardiomyocytes," *Nature* 485 (2012): 593–600.

Chapter 16: Managing a Catastrophic Success

246   **industrial raw material prices are increasing:** For example, on the London Metal Exchange a range of industrial metals (including copper, nickel, zinc, and steel) have doubled, tripled, or quadrupled in price in the last ten years.

246   **petroleum output has recently flattened:** As usual, however, new technologies make trends a treacherous guide to the future and views differ. For example, the International Energy Agency released a report this week (in 2012) that discusses unconventional oil technologies and states that, contrary to widespread expectations, "The United States is projected to be-come the largest global oil producer before 2020," becoming a net exporter around 2030. The IEA report then warns of a different constraint: the impact of CO_2 emissions ("World Energy Outlook 2012 Factsheet," International Energy Agency).

252   **CO_2 levels would remain high for centuries:** S. Solomona et al., "Irreversible Climate Change Due to Carbon Dioxide Emissions," *Proceedings of the National Academy of Sciences* 106, no. 6 (2009): 1704–1709 (http://www.pnas.org/content/106/6/1704.long).

252   **to lower CO_2 levels quickly and deeply:** And even this, by itself, may not be enough: "[T]he climate change that takes place due to increases in carbon dioxide concentration is largely irreversible for 1,000 years after emissions stop. Following cessation of emissions, removal of atmospheric carbon dioxide decreases radiative forcing, but is largely compensated by slower loss of heat to the ocean, so that atmospheric temperatures do not drop significantly for at least 1,000 years" (from S. Solomona et al.).

252   **collect and compress three trillion tons of CO2:** While the main difficulty is capture and compression to liquid densities, sequestration per se is another. At liquid density, the quantity in question (about three trillion tons) is on order of one-tenth the volume of Lake Baikal, or one-ten-thousandth the volume of the Antarctic ice cap—enormous volumes, yet small

on a global scale. Storage of pressurized CO_2 in geological formations on this scale seems questionable; storage in a granular solid form seems feasible, and at what will then be an affordable cost as measured by additional materials, energy, and processing.

Chapter 17: Security for an Unconventional Future

260   **aerospace systems, where APM-based production:** I discuss a more detailed scenario in "The Stealth Threat: An Interview with K. Eric Drexler," *Bulletin of the Atomic Scientists* 63, no. 1 (2007): 55–58.

265   **not only society as a whole, but also those who imagine that they can control it:** Power struggles have often placed those near the apex at greatest risk.

266   **pressures to compete for access to markets and natural resources:** In the United States, for example, federal documents describe competition for natural resources as a key concern in national security strategy:

From the Office of the President, "National Security Strategy" (2010): "America—like other nations—is dependent upon overseas markets to sell its exports and maintain access to scarce commodities and resources," and "as long as we are dependent on fossil fuels, we need to ensure the security and free flow of global energy resources."

From the Office of the Chairman of the Joint Chiefs of Staff, "National Military Strategy" (2011): "The persistent challenge of resource scarcity may overlap with territorial disputes."

From the Office of the Secretary of Defense, "Military Power of the People's Republic of China" (Annual Report to Congress, 2008): "As China's economy grows, dependence on secure access to markets and natural resources, particularly metals and fossil fuels, is becoming a more urgent influence on China's strategic behavior."

From the US Joint Forces Command, "Joint Operating Environment" (2010): "A severe energy crunch is inevitable without a massive expansion of production and refining capacity . . . an economic slowdown would exacerbate other unresolved tensions, push fragile and failing states further down the path toward collapse, and perhaps have serious economic impact on both China and India. . . . One should not forget that the Great Depression spawned a number of totalitarian regimes that sought economic prosperity for their nations by ruthless conquest."

269   **enduring importance of military strength and arms development:** There are several reasons for this. Going forward from the present, APM prospects will not quickly change the need to maintain strong force postures. Looking further, toward problems of transition management, changes in national interests will require strategies and forces suited to meeting the challenges discussed in the main text. Finally, the advent of APM-level technologies will entail a new revolution in military affairs. In that context, a cooperative framework for collective security will shape force requirements, but these requirements will surely include strong security systems based on advanced technologies.

270   **Cooperative Strategies for Avoiding Needless Risks:** Cooperative strategies for avoiding needless security risks are already on the agenda, in a general way. In the United States, for example, from the Office of the President, "National Security Strategy" (2010): "Disagreements should not prevent cooperation on issues of mutual interest, because a pragmatic and effective relationship between the United States and China is essential to address the major challenges of the 21st century." From the Office of the Chairman of the Joint Chiefs of Staff, "National Military Strategy" (2011): "Preventing wars is as important as winning them, and far less costly," a view that concurs with opinions throughout most of the world and its history.

Chapter 18: Changing Our Conversation About the Future

283   **the first response . . . often sets the direction for the next:** This is an example of a "social cascade," discussed (together with a range of other successes and pathologies of group decision-making) in Cass Sunstein's brief and readable book, *Infotopia: How Many Minds Produce Knowledge* (Oxford, UK: Oxford University Press, 2006).

Actin, 69
Additive manufacturing, 76–77
Agriculturalists, hunter-gatherers *vs.*,
    41–42
Agricultural Revolution, 39, 40–42
    APM Revolution and, 50, 54
    Industrial Revolution and, 44
    nature and human impacts of, 54
Agriculture, atomically precise
    manufacturing and, 231–232, 248,
    250
American Chemical Society, 181
Angewandte Chemie (journal), 20*n*
APM. *See* Atomically precise
    manufacturing (APM)
APM Revolution, 39, 40, 50–53, 54
    Agricultural Revolution and, 50, 54
    consequences, potential, 240, 286
    competitive, 243
    Information Revolution compared,
        xii, 256
    nanotechnology research and, 202
    nature and human impacts of, 54
    personal concerns and, 282
    threshold of, 193

Apollo program, 18, 20, 111–112
Applications, atomically precise
    manufacturing, 166–167, 174,
    223–239, 281
    consumer products, 224–225, 253
    security, 263–266
    medical, 167, 236–238, 256
    military, 35, 236, 259–263, 284
Approximations, exploiting, 123–124
Arms race, military applications of
        atomically precise manufacturing
        and, 259, 261–262, 268–269, 284
Armstrong, Neil, 112
Arrhenius equation, 292
Assemblers, 329
Assembly methods, molecular, 190–193
Asteroid mining, 15*n*
Astronautics, 133
Atomically precise fabrication, 177–193
    biological examples of, 80–82
    biomolecular engineering as, 182–184
    chemical synthesis as, 82–84, 179–181,
        182
    history of, 22–25, 46, 178
    materials science and, 184–185

Atomically precise fabrication *(continued)*
  nanotechnology as, 28, 195–196
  National Nanotechnology Initiative
    and, 32, 205, 207
  pathway to atomically precise
    manufacturing, 9, 25–27, 32–33,
    84–86, 144, 280–281
  scanning probe methods, 185–186
Atomic precision, x, xiii, 7, 10, 22–24, 50
  called the essence of nanotechnology,
    205
  definition of, 7
  digital systems compared to, 7, 77–80
  feedstock molecules and, 152–153
  from small to large scale, 154–155
  long history of, 22
  nanolithography compared to, 76
  nanotechnology and, x, xiii, 32
  Richard Feynman and, 24
  See also Atomically precise fabrication,
    Atomically precise manufacturing,
    Chemistry
Atomically precise manufacturing
    (APM), x–xii
  agriculture and, 231–232, 248, 250
  aligning national interests and,
    266–269
  applications (*see* Applications)
  assembly methods and, 190–192
  automated manufacturing as template
    for, 73–77, 84
  biomolecular engineering and,
    187–188
  biotechnology and, 73, 80–82, 85
  carbon dioxide and, 234, 246,
    250–252, 255
  chemical synthesis and, 73, 82–84, 85
  chemistry and, 179–181
  civil society and, 262–263, 265–266
  collaboration in, 271–272, 312
  consumer products from, 224–225,
    253
  costs and, 52, 224, 227
  digital information systems and, 73,
    77–80, 84
  digital media as cost model for,
    172–173
  digital revolution/digital technology
    and, 7–8, 50–51, 53, 277–278
  domestic security applications, 263–266
  economic implications of, 34–35,
    256–257
  energy and, 226, 229–230
  environmental restoration and, 33–34,
    233–234, 250–252, 255
  exploratory engineering and, 143–144,
    279–280
  framework for thinking about,
    274–287
  fundamental principles of, 10, 24,
    289–293
  improvement in product performance
    and, 162–166
  information technology and, 226–227
  materials processing and, 184–185
  medicine and, 236–238, 256
  military applications, 35, 236,
    259–263, 284
  molecular biology and, 24–27
  pace and direction of development,
    241–245, 309–311
  pathways to (*see* Pathways to
    atomically precise manufacturing)
  potential solutions/disruptions created
    by, 34–35, 240–241, 245–255
  precursors, 279, 303–306
  productivity of, 276–277
  progress towards, 32–33, 177–179, 278
  raw materials and, 230–231
  relationship to nanotechnology, 32,
    196–199, 205–207
  reducing complexity of, 303–304
  resource scarcity and, 33–34, 169,
    230–231, 248
  roadmapping for progress in, 216–220
  scanning-probe methods and, 185–186
  security technologies and, 235–236
  supply chains and, 34–35, 51, 225–226,
    244–245
  surveillance networks and, 263,
    264–266
  transformation of infrastructure and,
    228–229
  uncertainties and, 258, 269–272
  See also Atomically precise
    manufacturing systems, Pathways
    to atomically precise
    manufacturing

Atomically precise manufacturing
(APM) research
carbon-based supermaterials and, 158
fostering collaborative strategies for,
271–272, 312
government funding and, 32, 194–195,
198–199, 204–207, 208, 243
repression of, 209–210
Atomically precise manufacturing
(APM) systems
energy requirements of, 155–156
as factories, 276
microblocks, 152–155
ordinariness of, 70–71
process of, 148–151
products of, 147–149, 159–174,
224–225, 253 (see also Applications)
radical cost reduction and, 168–173
Automated manufacturing, APM and,
72–77, 84
Avco Everett Research Laboratory, 17–18

Battelle Memorial Institute, 211
Becquerel, Henri, 134
Bell Telephone/Bell Labs, 46–47
Biomolecular
engineering, 9, 187–188
machine engineering, 24–25
systems, atomically precise
manufacturing and, 73, 80–82, 85
Boltzmann factor, 292n
Brownian motion, 23, 50
Bush, Vannevar, 5

CAD (computer-aided design) software,
189
CAMD (computer-aided molecular
design) software, 189–190
Cancer, atomically precise manufacturing
and attack on, 237–238
Carbon-based materials, 137, 153, 158,
162–163
Carbon dioxide
emission reduction, 171, 250–252, 255
problem, 246, 250–252
removal from atmosphere, 234, 252,
255
See also Greenhouse gases
Carbon nanotubes, 161, 164, 185, 188

Carroll, Sean, 96, 100
Carson, Rachel, 12
Casimir forces, 64
Catalysts (as products), 302
Cells, molecular machinery of, 25, 50,
61–62, 182
CERN, 95
Chemical synthesis
atomically precise manufacturing and,
73, 82–84, 85
organic synthesis, 23–24, 32, 179–181,
187
Chemical reactions
click chemistry/click reaction, 180n
equilibria of, 292
free energy change, 291–292
in chemical synthesis, 23, 83, 84, 180,
293
kinetics of, 277, 292
machine-guided motion, 73, 292
methods for blocking, 84, 281,
290–292, 300
stereotactic, 290–293
thermal motion and, 68
thermal motion timescale, 68
thermodynamic control of, 277, 292
transition states of, 291, 292
yield, 84, 113, 292
See also Chemical synthesis,
Chemistry
Chemistry
atomically precision and, 7, 22–23,
82–84
computational, 33, 56, 98–100, 179,
189–190, 218
discovery of atoms and, 23, 29
organic synthesis and, 179–181
as pathway technology, 32, 84–85,
179–181, 188, 242
as a production method, 92–84
research scope, 178–181
thermal motion and, 68–70
China
conflict with United States, 268–269
economic rise of, 246
government funding of
nanotechnology research, 194, 210,
243
Churchill, Winston, 40

Civil society, atomically precise manufacturing and, 262–263, 265–266

Clausewitz, Carl von, 262

Climate change, atomically precise manufacturing and, 234, 246, 250–252, 255. *See also* Carbon dioxide, Greenhouse gases

Clinton, Bill, 194, 204, 207

Collaboration in scientific research, 271–272, 312

Competition, effect on engineering, 140

Computational chemistry, 33, 60, 98–100, 157, 274*n*

Computational systems, atomically precise manufacturing and, 165–166, 167, 218

Computational tools for design, 189–190

Computers, history of, 47–48

Congress, United States
molecular manufacturing study and, 210–211
National Nanotechnology Initiative and, 204, 205–206, 208

Construction, APM-based production and, 228

Consumer economy, globalized, 51

Consumer products, atomically precise manufacturing and, 224–225, 253. *See also* Applications, Products

Control signals, 227

Cooperative strategies, to avoid risk, 270–272

Costs, atomically precise manufacturing and, 157, 168–174, 224–231, 226, 227, 244–245, 239, 326
digital media model, 172–173
energy and, 170
industrial accidents and, 171
labor, directly involved, 168, 170
land and, 170
non-physical costs, 168–169, 172–173
physical capital, 171–172, 228, 230–232, 248
physical cost concept, 168, 169
radical reductions in, 168
raw materials and, 169–170
total, 172
wastes, 156, 171, 225, 245, 250, 282

Covalent solids, 297

Crick, Francis, 12, 24

Cross-linking, 298

Crystal engineering, 102–103

Dalton, John, 23

Darwin, Charles, 209

Democritus, 23

Department of Energy, 16

Desalination, atomically precise manufacturing and, 231, 248

Design
accelerating, 307–308
atomically precise manufacturing and cost of, 173
computational tools for, 189–190
open-source, 308–312
self-assembly and, 191
systems engineering and, 118–121
theories *vs.*, 117–118, 128–129

Digital media, as cost model for atomically precise manufacturing, 172–173

Digital principle, 78–79

Digital revolution/digital technology, xii, 3–7
atomically precise manufacturing and, 7–8, 50–51, 53, 73, 77–80, 84, 277–278
nature and human impacts of, 54

Dirac, Paul, 98

Dispersion forces, 64

DiVincenzo promise criteria, 216

DNA, 80, 81
as digital instructions, 182
structure of, 24
structural DNA engineering, 81, 182, 184, 187–188, 308
synthetic, 181

Drones, 236, 261, 262, 264

DuPont, 46

Earth-abundant materials, atomically precise manufacturing and, 166, 169, 230

Earth Day, 12

Economic implications, of atomically precise manufacturing, 34–35, 256–257, 282

asset revaluation, 254–255
disruption, 34, 241, 254, 282
employment, 341, 254 (*see also* Labor)
market access, 266–267
trade, 34–35, 266, 282
supply chains, 34–35, 51, 225–226,
    244–245
Edison, Thomas, 46
Eigler, Don, 185
Einstein, Albert, 93, 95
Electric motors, nanoscale, 63, 155,
    164–165
Electronic phenomena, unusual, 97, 115,
    164
Energy
    atomically precise manufacturing and
        cost of energy inputs, 170, 172
    atomically precise manufacturing and
        energy infrastructure, 226, 229–230
    requirements for atomically precise
        manufacturing, 155–156
    scaling principles and, 164–165
Engineering
    applying perspective of, 130–131
    approach to knowledge, 121–123
    within bounds of limited knowledge,
        103–105
    cost and performance of products *vs.*
        design, 141–143
    crystal, 102–103
    design and, 113–115, 117–118, 122–130
    effect of competition on, 140
    exploiting approximations in, 123–124
    exploratory (*see* Exploratory
        engineering)
    goal-oriented development in,
        129–130
    molecular systems (*see* Molecular
        systems engineering)
    parable of the blind men and the
        elephant and, 109–111
    protein, 25–27, 29–30, 32, 182–184,
        187, 212
    reliability and, 7, 124–125, 137, 141
    science-intensive, 217
    science *vs.*, 28–30, 104–105, 109–115,
        121–130
    structural DNA, 81, 182, 184, 187–188,
        308

system-level predictability and, 124,
    125
systems level design, 118–121
use of scientific knowledge in, 92–93,
    94
See also Systems engineering
Engines of Creation (Drexler), 10, 27, 30,
    31, 194, 197, 200, 202, 207, 209
Enthusiasm as a problem, 275
Environmental movement, 12–13, 15
Environmental restoration, atomically
    precise manufacturing and, 33–34,
    233–234, 249–252, 255
Enzymes, 24, 81–82, 183
Exosomes, 167
"The Exploration of Cosmic Space by
    Means of Reaction Devices"
    (Tsiolkovsky), 133
Exploratory engineering
    applied to molecular systems
        engineering, 218
    atomically precise manufacturing and,
        143–144, 279–280
    competition and, 140
    cost and performance of design and
        analysis, 141–143
    knowledge as product of, 137–143
    rules of, 139–140
    science *vs.*, 132
    spaceflight and, 132–136
    systems engineering and, 136–137
    See also Landscape of technological
        potential

Feedstocks, for atomically precise
    manufacturing, 153–154
Feynman, Richard, 24, 98, 194
Fischer, Hermann Emil, 24
Fish, nanoscale structure of, 61
Fishing line, nanoscale structure of,
    60–61
Fluids, atomically precise manufacturing
    and, 300–301
Foldamers, 181, 183–184, 188, 296, 302
Foldit, 308*n*
Fold prediction/design, 29, 103, 308*n*
Food production, potential impact of
    atomically precise manufacturing
    on, 231–232

Friction, 65–67. *See also* Superlubricity
Fuel cells, 165

Gasoline engine, nanoscale structure of,
    62–63
General Electric, 46
Genetic engineering, 25–26, 182
Genome research, 112–113, 237
Gift economy, 256–257
Goddard, Robert, 135
Gore, Al, 31
Government funding of nanotechnology
    research, 32, 194–195, 198–199,
    243
    National Nanotechnology Initiative,
        204–207, 208
Government support for industrial
    research and development, 46
Greenhouse gases, 171, 246, 250. *See also*
    Carbon dioxide
Graphene nanostructures, 305
Gray goo, 201*n*, 207

Haber, Fritz, 44
Hapgood, Fred, 27
Heisenberg, Werner, 98
Higgs boson, 95
History
    of atomically precise fabrication,
        22–25, 46, 178
    of idea of atomically precise
        manufacturing, 24–25, 27–32,
        194–211
    of nanotechnology research programs,
        204–210
Hoffmann, Roald, 179
Human development, atomically precise
    manufacturing and, 246–248
Human Genome Project, 112–113
Hunter-gatherers, agriculturalists *vs.*,
    41–42

India
    economic rise of, 246
    nanotechnology research in, 243
Industrial civilization, crisis confronting,
    245–246
Industrial equipment, APM-based
    production and, 228

Industrial production
    atomically precise manufacturing and,
        ix–x, xii
    experimental devices *vs.*, 161
    Information Revolution and, 48–49
Industrial civilization, crisis of limits,
    245–246
Industrial Revolution, 39, 40, 42–47, 54
    APM Revolution and, 50–51, 53, 54
    metrics for progress during, 295
Infection, atomically precise
    manufacturing and attack on,
        237–238
Information Revolution, 6–7, 39, 40, 47–49
    APM Revolution and, 50–51, 53
Information technologies, atomically
    precise manufacturing and,
        226–227
Insulating materials, thermal 228
Intel, 48
Interagency Working Group on
    Nanoscience, Engineering and
    Technology, 204
*International Technology Roadmap for
    Semiconductors* (ITRS), 215
Internet, 47, 49, 53, 224, 245

Joy, Bill, 207, 208

Kalam, Abdul, 194
Kantrowitz, Adrian, 18
Kantrowitz, Arthur, 17–19
Kawczak, Alex, 211
Kendrew, John, 24
Kennedy, John F., 20
Knowledge
    exploratory engineering and, 134,
        137–143
    scientific, 92–94
    seeking *vs.* applying, 121–123
    specialization of, 45–46
Korolev, Sergei, P., 12
Krupp, Alfred, 46

Labor, 51, 168, 170
    agricultural, 41–42
    atomically precise manufacturing and,
        51, 170, 241, 245, 256, 282
    division of, 46

industrial, 44, 51–52
information and, 51
skills, 4
Land, atomically precise manufacturing
and cost of, 170
Landscape of technological potential,
89–92, 94–95, 284–285
atomically precise manufacturing and,
91, 95, 105–106, 143–144
exploratory engineering and, 131, 132,
137
limits of, 91–92
partially visible, 92, 94, 97, 101, 131
technology development and, 92,
143–144, 279
timeless, 91–92, 94, 174, 240, 284–285
Lane, Neal, 206
Lightsails, 17, 20
Limits to Growth (Meadows), 13, 246
Limits to growth, 10–11, 15–17, 41, 53,
245–246
The Logic of Scientific Discovery (Popper),
126
Lucretius, 23

Macromolecular structures, 218
systems engineering and, 187–188,
243
foldamers, 181, 183–184, 188, 296, 302
See also Protein molecules, DNA
Macromolecules, folded, 296
Marconi, Guglielmo, 134
Markets, reducing competition for access
to, 266–267
The Mars Project (Das Marsprojekt) (von
Braun), 135
Material economic development,
atomically precise manufacturing
and, 246–248
Materials
for atomically precise manufacturing,
144, 160–161
atomically precise manufacturing and
demand for raw, 230–231
gradient of quality for, 296–298
improvement in performance of,
162–166, 228
interaction in nanoscale world, 64–65
processing, 184–185

synthesized by atomically precise
manufacturing, 163–164
Medical applications, of atomically
precise manufacturing, 167,
236–238, 256
diagnosis and, 237
Medvedev, Dmitry, 194
Metallic bonding, 62
Metallurgy, 45–46
Metal-organic frameworks, 298
Metals, atomically precise manufacturing
and, 169
Metrics, for progress, 294–295
Microblocks, 151–152, 302–303
construction of, 152–155
Military applications of atomically
precise manufacturing, 35, 236,
259–263, 284
Mining
atomically precise manufacturing and,
34, 231
environmental restoration and, 233
of space resources, 15*n*, 16
MIT Space Habitat Study Group, 17
Molecular biology, 61–62. *See also*
Biomolecular
atomically precise manufacturing
compared, 24–27
thermal motion and, 68–70
Molecular dynamics, 60, 274*n*
"Molecular Engineering" (Drexler), 8*n*,
196*n*
Molecular-level physical principles of
atomically precise manufacturing,
289–293
Molecular machines, 24–27, 61
Molecular manufacturing, 209
government study on, 210–211
See also Atomically precise
manufacturing (APM)
Molecular mechanics, 33
Molecular systems engineering, 187–188,
243
current state of, 217
road ahead, 218
Molecules
atomic structure of, 22–23
as atomically precise, 7
chemical synthesis of, 179–181

Molecules *(continued)*
    construction of microblocks from,
        152–155
    mechanical characteristics of, 56–57
    small, as feedstocks, 153
    structure of, 22–24, 55–56
Monomers, 297
    activated, in atomically precise
        manufacturing 153, 154–155
    DNA and, 181
    foldamer diversity and, 183–184
    peptides and, 181
Mineral resources
    Earth-abundant, 166, 169, 230
    scarce, 230, 248, 250
Moon
    space settlement and, 15*n*
    See also Apollo program
Moore, Gordon, 48
Moore's Law, 48, 75, 311*n*
    roadmapping, 215
Morrison, Philip, 14
Music, digital revolution and, 3–4
Myosin, 69

Nanobots, 208–209
Nanobugs, 201*n*, 203–204, 208
NanoEngineer-1, 308
Nanolithography, 75–76
Nanomachines, 57, 59, 63–71, 125, 203
    biomolecular, 69–70, 80–82, 125
    conventional nanotechnology and,
        185, 198, 202–203, 205
    friction and, 65–67
    mistaken for nanobugs, 28
    ordinariness of, 70–71
    surfaces, texture of, 63–65
    thermal motion and, 67–70
    See also Atomically precise
        manufacturing (APM), molecular
        machines
Nanomedicine, 237–238
Nanoparticles, 161, 185
Nanorex, 308
Nanoscience, 28, 115
Nanosystems: Molecular Machinery,
    Manufacturing, and Computation,
    30, 194, 211

Nanotechnology
    as atomically precise fabrication,
        195–196
    as brand name for different kind of
        research, 28, 199, 203, 278–279
    change in concept of, xi, xiii, 27–28,
        30–31, 197–200
    conflict between popular vision and
        scientific reality of, 31–33, 207–209
    definitions of, 205, 206
    features of, x–xi
    government funding of, 32, 194–195,
        198–199, 204–207, 208, 243
    nanoscience and, 115
    popular focus on tiny robots, 199–200,
        201
    origins of idea of, 9–21
    redefined as nanoscale material and
        devices, 196
    research, 202–204
Nanotechnology: Shaping the World
    Atom by Atom (brochure), 204,
    206
Nanotubes, 161, 164, 185, 188
NASA, 15, 16
    Ames Research Center, 209
    Jet Propulsion Laboratory, 17
    Langley Research Center, 18
National Academy of Sciences, 210, 211
National interests, reassessing, 174, 261,
    266–269, 284
National Nanotechnology Initiative: The
    Initiative and its Implementation
    Plan, 204–205
National Nanotechnology Initiative
    (NNI), 204–207, 208–210
National Nanotechnology Initiative
    Strategic Plan, 206
National Science and Technology
    Council (NSTC), 204
National Science Foundation, 46
Nature (journal), 20*n*, 100
Neolithic Revolution, 40–42, 54
Newton, Isaac, 127
Newtonian mechanics, 93, 98–99, 134
New York Times (newspaper), 18
Nicholas I, 133
Nitrogen fertilizers, 44

NNI. *See* National Nanotechnology Initiative (NNI)
Noise margins, 7–8, 79
Non-lethal force, atomically precise manufacturing and, 236, 261–262
Non-state actors, constraining, 267
NSTC. *See* National Science and Technology Council (NSTC)
Nylon, 60–61, 181

Oak Ridge National Laboratory, 209
Oberth, Hermann, 135
OMNI (magazine), 27
O'Neill, Gerard K., 14–15
Open-source model, atomically precise manufacturing and, 173, 308–312
Oregon State University, 11
Oxides, 160, 169, 297

Parable of the blind man and the elephant, 107–113
Pathways to atomically precise manufacturing, 281, 294–312
  atomically precise technology gradient, 296
  estimating pace of progress, 309–312
  gradient in environments and means of transport, 299–301
  gradient of complexity, 301–303
  gradient of levels of stereotactic control, 298–299
  gradient of quality for materials and components, 296–298
  looking backward, 303–306
  metrics for progress, 294–295
  open-source design and, 308–312
  pace of development, 306–308
Peptides, synthesis of, 181
Peptoids, 181, 184, 188, 189, 296
Performance, design and, 141–143
Perutz, Max, 24
Photovoltaic cells, atomically precise manufacturing and, 164, 166–167, 302
Physical laws
  atomically precise manufacturing and, 105–106

potential technologies and, 9–10, 25, 91–92, 105–106
  systems engineering and, 118–119
  technological implications of, 137
  universality and limitations of, 94–97
  use of, 92–94
Physics
  engineering within bounds of, 103–105
  limits of knowledge of, 94–97
  of technology, 90–92
  textbook, 92–94
  unknown and unpredictable and, 100–103
Pollution, atomically precise manufacturing and cost of, 171
Polymers, 296–297. *See also* Proteins, DNA
Polypeptides, 296. *See also* Proteins
Popper, Karl, 19, 126
Population growth, food production and, 40–42, 44, 231
POSCO, 52
Poverty and economic development, 168, 243, 247–248
Precursors, of atomically precise manufacturing, 279, 303–306
Printing, 4
  3D, 76–77
Proceedings of the National Academy of Sciences (journal), xiv, 27n
Production
  atomically precise manufacturing and the transformation of, 225–226
  atomically precise manufacturing and capacity, 253–254, 265
  changes in concept of, 49
  cost of atomically precise manufacturing, 157
  food, 231–232
  industrial (*see* Industrial production)
Productive Nanosystems: A Technology Roadmap, 216
Productivity, of atomically precise manufacturing, 54, 276–277
  scaling laws and, 75, 84, 149–150, 154, 225
  See also Costs, atomically precise manufacturing and

Products
    atomically precise manufacturing,
        147–149, 159–174
    atomically precise manufacturing and
        improvements in performance,
        162–166
    consumer, 224–225, 253
    cost and performance of, *vs.* design
        and analysis, 141–143
    cost of atomically precise
        manufacturing, 141–143, 168–173
    development cycle, 244–245
    engineering reliable, 124–125, 141
    improved performance through
        scaling, 164–166
Proteins
    engineering of, 25–27, 29–30, 32,
        182–184, 187, 212
    fold prediction, 29, 103, 308n
    as materials, 26, 296
    as objects, 24, 26, 69, 297
    as molecular machines, 26, 69–70
    motion of, 69
    self-assembly and, 26

Quantum chemistry, 33, 157
Quantum dots, 185, 188
Quantum Information Science and
    Technology Roadmap (QISTR),
    215–216
Quantum physics, 46–47, 93–94, 98

Radical abundance, 286
    aspects of, 52
    digital revolution and, xii, 3, 6–7, 49
    emergence of, 274
    enabling technology, 147–158
    products of, 159–174
    scaling laws and, 59
Range of possibility, science and
    engineering and expanding, 125–126
Ralph Merkle, 321
Raw materials, atomically precise
    manufacturing and, 169–170, 172,
    230–231
Re-entry, atmospheric 18, 20
Regenerative medicine, 238
Relativity, 95, 96

RepRap community, 77
Research and development, Industrial
    Revolution and systematic, 46
Resilience, afforded by atomically precise
    manufacturing, 248–249
Resources, material and energy, 246
    common, atomically precise
        manufacturing and 169–170, 230
    demand, atomically precise
        manufacturing and, 248
    prices, atomically precise
        manufacturing and, 255
    reducing international competition
        for, 266–267, 267–268
    scarcity, atomically precise
        manufacturing and, 33–34, 169,
        230–231, 248
Ribosomes, 81, 183, 302
Risks
    arms and, 35, 236, 257, 259–260, 267
    cooperative strategies for avoiding
        needless, 270–272
    disruptive change, 243
    industrial accidents and, 171
    popular misconceptions of, 201, 208
    uncertainty and, 240, 261, 269–270
RNA, 70, 81, 237
Roadmapping
    Moore's law and, 215
    for progress in atomically precise
        manufacturing, 211, 216–220
    quantum information systems,
        215–216
    in semiconductor industry, 214–216
Robots, nanotechnology and focus on
    tiny, 199–200, 201, 208–209
Rocket technology, 133–135
Röntgen, Wilhelm, 134

Safety margins, 105, 141, 142
Scaling laws, 56, 58, 63n, 149–150, 154,
    162, 163, 165, 289, 299
    atomically precise manufacturing and,
        75
    improved product performance
        through, 164–166
    time, 58–59
Scanning-probe techniques, 32, 185–186

Schrödinger's equation, 98
Schwinger, Julian, 98
Science
    approach to knowledge, 121–123
    curiosity-driven investigation in,
        129–130
    engineering vs., 28–30, 104–105,
        108–115, 121–130
    Industrial Revolution and growth of,
        45
    inquiry and, 113–116, 121–130
    natural phenomena and, 124–125
    parable of the blind man and the
        elephant and, 108–111
    seeking precision in, 123–124
    simplicity of theories in, 128–129
    testing theories in, 126–127
Science-intensive engineering, 129, 180,
    213–214, 217
Science (journal), 20n, 100, 198n
Scientific American (journal), 208
Scientific research, collaboration in,
    271–272, 312
Scientific Review (journal), 133
Security technologies, atomically precise
    manufacturing and, 235–236,
    259–266
Seeman, Nadrian, 184
Self-assembly, 144, 186–188, 190–193,
    218, 293, 297, 299–300
    thermal motion and, 71, 300
Self-replicating machines, 201. See also
    Nanobots; Nanobugs; Smalley,
    Richard
Semiconductor fabrication, 75–76, 311n
    atomically precise manufacturing and,
    226
Semiconductor industry, roadmapping
    and, 214–216
Sensors, 227, 235
Sharpless, Barry, 180
Silent Spring (Carson), 12
Silicates, 301
Smalley, Richard, 208–209
Smith, Adam, 45
Solar energy, atomically precise
    manufacturing and, 34, 164, 230,
    233, 248

Soviet Union, spaceflight and, 133–135
Space development, 11–15, 18–19
Spaceflight, history of, 132–136
Space industrialization, 15–17
Space settlement, 15–16
Space systems engineering community,
    19–21
Sputnik, 12, 133
Standard Handbook of Machine Design,
    140
Standard Model, 95–96, 128
Stereotactic control (molecular),
    190–193, 214, 218, 290–293,
    300–301
    atomic precision through, 290–291
    broad applicability of, 292–293
    defined, 190
    design and, 307, 311
    effects of, 311–312
    fluids and, 300
    gradient of levels of, 298–299
    physical requirements of, 291–292
    reaction rates, 291
    self-assembly and 292–293, 298–299
    structural stability and, 291
    synthesis/assembly, 190, 191–193
    thermal motion and, 191, 277,
        292–293, 299–301
Stiffness, 67–68, 70, 298, 299
Structural DNA engineering, 81, 182,
    184, 187–188, 308
Structural DNA nanotechnology, 33, 212,
    297
Sun Microsystems, 207
Sun Tzu, 261–262
Surfaces
    of machine components, 63–67, 122,
        127
    of protein molecules, 69
    self assembly and, 144, 297
Superlubricity, 66
Supply chains, effect of atomically
    precise manufacturing on, 34–35,
    51, 225
Surveillance, 227, 235–236, 263–266
Sustainable development, 248–249, 282
System-level predictability, science and
    engineering and, 124–125

Systems engineering, 118–121, 126–127, 130
    exploratory engineering and, 136–137
    macromolecular, 121, 187, 189, 196, 214, 217–218
    space, 19–20, 113–114, 133–135

Terrorist attacks, domestic security and, 264–266
"There's Plenty of Room at the Bottom" (Feynman), 24, 194
Thermal motion, 7–8, 23, 60–61, 67–70, 102–103, 191, 277, 291–293, 299–301
    in atomically precise manufacturing, 79, 277, 291
    machines and mechanical restraints, 7–8, 67–68, 277, 279
    in molecular biology, 61, 63, 69–70, 81, 103
3-D printing, 76–77
Thermodynamics
    limits and, 156, 167, 228, 234, 277
    chemical reactions and, 156, 292
Tomonaga, Sin-Itiro, 98
Toxicological hazards, atomically precise manufacturing and, 161, 250
Tsiolkovsky, Konstantin, 133–135

United States
    conflict with China, 268–269
    government funding of nanotechnology research, 194–195, 204–207, 209–210, 243
    See also U.S. Congress
U.S. National Laboratories, 46, 211, 216
Uncertainty, knowledge of, 20, 126, 240, 258, 269–270
Universe, beginnings of, 89

Vacuum aerogels, 228
van der Waals forces, 64
Verne, Jules, 133
von Braun, Werner, 135

Waste, atomically precise manufacturing and, 156, 171
Water scarcity, 231–232, 246
    atomically precise manufacturing and, 231, 248, 300
Watson, James, 12, 24
Weapons, non-lethal, 261–262. See also Military applications of atomically precise manufacturing
Westinghouse, 46
WiFi, 227
Wikipedia, 49
Wilkins, Maurice, 24
Wired magazine, 207

Zyvex stereotactic method, 300

K. Eric Drexler developed and named the concepts of nanotechnology and atomically precise manufacturing, and brought them to the attention of a broad audience. Currently with the Oxford Martin School at Oxford University, Drexler is a frequent public speaker on science, technology, and policy, addressing audiences of business and government leaders, engineers, and scientists in the Americas, Europe, and Asia.

PublicAffairs is a publishing house founded in 1997. It is a tribute to the standards, values, and flair of three persons who have served as mentors to countless reporters, writers, editors, and book people of all kinds, including me.

I. F. STONE, proprietor of *I. F. Stone's Weekly*, combined a commitment to the First Amendment with entrepreneurial zeal and reporting skill and became one of the great independent journalists in American history. At the age of eighty, Izzy published *The Trial of Socrates*, which was a national bestseller. He wrote the book after he taught himself ancient Greek.

BENJAMIN C. BRADLEE was for nearly thirty years the charismatic editorial leader of *The Washington Post*. It was Ben who gave the *Post* the range and courage to pursue such historic issues as Watergate. He supported his reporters with a tenacity that made them fearless and it is no accident that so many became authors of influential, best-selling books.

ROBERT L. BERNSTEIN, the chief executive of Random House for more than a quarter century, guided one of the nation's premier publishing houses. Bob was personally responsible for many books of political dissent and argument that challenged tyranny around the globe. He is also the founder and longtime chair of Human Rights Watch, one of the most respected human rights organizations in the world.

· · ·

For fifty years, the banner of Public Affairs Press was carried by its owner Morris B. Schnapper, who published Gandhi, Nasser, Toynbee, Truman, and about 1,500 other authors. In 1983, Schnapper was described by *The Washington Post* as "a redoubtable gadfly." His legacy will endure in the books to come.

Peter Osnos, *Founder and Editor-at-Large*